VEHICLE BODY LAYOUT AND ANALYSIS

BY

JOHN FENTON

MECHANICAL ENGINEERING PUBLICATIONS LTD
LONDON

FIRST PUBLISHED 1980

ISBN 0 85298 445 6

Filmset by Santype International Limited, Salisbury, Wilts.

Printed by St. Edmundsbury Press Limited, Bury St. Edmunds, Suffolk.

VEHICLE BODY
LAYOUT AND ANALYSIS

To Paul, Peter, and Denise

CONTENTS
and
STEPS TO DESIGN ANALYSIS

PREFACE

THE FUEL crisis is directing the attention of automotive engineers to reduce vehicle weight. The fundamental drag force on road vehicles at low speed is tyre rolling resistance. It is directly proportional to the weight imposed through the road wheels on to the ground. Gradient resistance too, is proportional to weight, as is the inertia force to overcome, on acceleration of the vehicle.

Vehicle manufacturers and equipment suppliers have hitherto invested large sums of money into reducing fuel consumption by carburettor and injection-system development, sophisticated ignition systems, and turbo-charging the air input to the engine. These efforts are tending to yield diminishing returns, with the passage of time, and the sophistication and expense of the equipment, and not least the manufacturing complexity.

However efficient the means of converting energy of the fuel into tractive force at the wheel-to-ground contact area, the fuel itself has a limited energy capacity. The energy contained in one gallon of refined petrol is typically 69 300 centigrade heat units, and engine thermal efficiencies are of 25 per cent at the most. So the 'potential' energy of the fuel, if equated to the kinetic energy (again proportional to weight) of a vehicle, moving at a given speed, reveals a finite energy value which no fuel-economizing device can ever exceed.

The case for reducing fuel consumption by reducing mechanical component and equipment weight, in turn reducing translation and rotational inertial drag on acceleration, as well as rolling resistance and gradient resistance, has been well argued. The configuration and assembly of components and structural elements was also of importance from a weight reduction standpoint. For example, a front wheel drive vehicle arrangement could save the weight of propeller shafting and drive axle casings. But unless introduced at the conceptual stage of a new vehicle the changeover from rear to front drive, after production tooling was commissioned, could be very expensive.

As excess weight at one part of a vehicle has to be supported by extra weight added elsewhere, conversely a weight reduction in one part may result in a meaningful saving in others. Also the fitting of a lighter system (such as a smaller engine) in one place could result in a mechanical change in another (smaller brakes) so there is a chain reaction—a lighter body structure may result in smaller wheels, and so on. The overall system design is thus of major importance.

A quantitative measure of the effect of weight saving on fuel economy is indicative in studies carried out to investigate automatic gearboxes by a UK manufacturer. These showed that the proportions of fuel used to accelerate the mass of a typical European car, when driven according to the USA Environmental Protection Agency 'Highway', 'Urban', and 'EEC 15' cycles, were 18, 38, and 25 per cent, respectively. Thus a one-third reduction in vehicle weight may be expected to yield some 6–12 per cent fuel economy, it was deduced. This comparatively modest yield meant that new component development for lighter weight was normally accompanied by design for lower cost in material and labour.

Vehicle weight also has a close relationship with body size so compact vehicles reap rich rewards in fuel savings, not only in weight saving but with proportionally reduced air drag as cross-section area is reduced. Now engineers are looking at the body structures themselves to see if the weight in them is efficiently distributed to react applied loads with the minimum of redundancy, to reduce 'skin' drag by smoothing surfaces and to provide passenger protection in accidents.

It has been stated recently by a prominent UK body engineer, that by applying computer-based finite-element structural analysis techniques to the design of new bodies, some 15 per cent weight saving could be achieved over the equivalent superseded model which had not had the benefit of these techniques. In this book the analytical techniques are summarized which are necessary to design the structure, as a preliminary stage, prior to finite-element analysis.

This introduction to vehicle structures will appeal to three main categories of reader; the vehicle body draughtsman and designer whose training may well have been restricted to a mechanical engineering course; the student trainee bent on vehicle engineering and reading the subject 'theory of structures' within a specialist engineering course as well as the specialist vehicle builder who will benefit from analytical techniques for obtaining greater structural efficiency in designing for low weight.

The analytical processes within the text will be of particular value to students in course levels between Technical T-5 and second year degree in vehicle engineering. Parts of the book will further be of value in the final year degree and post graduate studies in vehicle engineering.

In one aspect at least the subject matter of vehicle body structures is comparatively new: that is the analysis of vehicle bodies built by modern manufacturing techniques. Such techniques have superseded the methods used by skilled coachbuilders who formerly built car, bus, and truck bodies in a way not far removed from that applying to the days of horse traction.

Until recent years students and apprentices in automotive engineering still entered the industry with a background of mechanical engineering training in such subjects as thermodynamics, theory of machines, hydraulics, mathematics, mechanical design, and strength of materials. The teaching of theory of structures had been more common in courses of aeronautical engineering, civil (structural) engineering, and naval architecture.

Within the vehicle manufacturing industry other trends are throwing a new emphasis on body structural design: these intensify the need for wider training in structural theory for body draughtsmen and designers. Automotive power units, transmission, and suspension systems are becoming perfected and standardized in a way that precludes extensive change with the introduction of new models. New models are much more to do with changes in taste and fashion affecting the shape of the body, and with designing for specific market sectors where bodywork is tailored to meet special requirements. At the same time there are strong commercial pressures to shorten the lead times of new model introductions and to meet the projected standards of structural safety without resort to extensive road proving programmes.

The increasing application of analytical techniques to the solution of these problems demands a greater knowledge of structural theory from those about to enter automotive engineering and from those already employed in vehicle body design and manufacture. The third category of reader, the specialist vehicle builder will find much to attract him in the ingenuity of the techniques of structural analysis. The text that follows does not take in the complete vehicle structural analysis by computer methods. Rather it is a study in the behaviour of elements which go to make up a vehicle body so that a 'structural feel' can be engendered, all important in the preliminary stages of vehicle design.

For those without previous engineering knowledge, reading an elementary text in Strength of Materials would give an understanding of basic stress-strain relationships, bending and torsional behaviour of simple solid-sectioned members under load and help with the assimilation of the theory developed hereafter.

The book is an *introduction* to vehicle structures in the sense of providing readers with an insight into analytical methods given in a wide variety of published sources such as technical journals, conference papers, and proceedings of engineering institutions. The analyses are therefore not necessarily fully developed or rigorously evaluated; recourse to the original references is necessary particularly in order to understand the limiting assumptions on which the analyses are based.

JOHN FENTON

GLOSSARY OF TERMS AND ABBREVIATIONS

ASAE Advanced School of Automobile Engineers, Cranfield (now School of Automotive Studies)

Anticlastic bending Complementary bending of a plate or beam at right angles to the primary bending due to the external loading

Asymmetric bending Of a vehicle structure due to one wheel of an axle striking a bump

Boom End load carrying member at the edge of one, or corner of two intersecting, panel(s)

Bulkhead Self supporting ring or fabricated plate sub-structure comprising a principal transverse slice of the main structure

Column-buckling Overall buckling of member, under end loading, into a single wave or the first few harmonics

Compliance Flexibility introduced into suspension anchorage to isolate high frequency vibration

Creep modulus Slope of tangent to stress–strain curve after a given time of loading

Cruciform bracing Use of two beams in bending (in a cross formation) to reinforce a rectangular frame in torsion

Elastic modulus Direct stress divided by direct strain

Finite element method Structural analysis of a highly redundant structure based on the behaviour under stress of small finite elements expressed simultaneously

Flange Edge end load carrying element of beam in bending

Flexural axis Line joining locus of shear centres of sections along a beam

Flexural stiffness Rigidity in bending

Form factor Ratio of elastic to plastic bending moment in beam

Free-free Support conditions for a beam held at intermediate span with ends free to move

Inst. Mech. Engrs Institution of Mechanical Engineers

Isotropic Elastic properties the same about the three perpendicular axes

Kink strut Member joining the flanges at the intersection of beams at an angle to one another (kinked)
Longeron Longitudinal member connecting the transverse bulkheads
Lozenging Diagonal deflection of panel or frame
Percentage proof stress Stress at a given percentage strain after yield
Pillarless saloon Without B-C post (centre pillar) in sideframe
Pitch motion In the vertical and longitudinal plane of the vehicle about a transverse axis
Plastic hinge Localized bending failure at yield on a beam or frame causing a hinged deformation of adjacent members
Poisson's ratio Ratio of lateral strain to longitudinal strain
Pressbrake Machine for folding metal in a single plane
Racking strains Deflections caused by dynamic twist of structure
Reach Diagram Plot of arm or leg articulation providing an envelope of comfortable reach dimensions for control operation
Roll motion In the vertical and transverse plane of the vehicle about a longitudinal axis
Sandwich panel Panel made up of high density rigid skins sandwiching a low density core material
SAE Society of Automotive Engineers (USA)
Section modulus Section second moment of area divided by distance from neutral axis to extreme stressed fibre
Shear centre Point through which external load on a beam section must pass to cause no twist of the beam
Shell structure Three-dimensional system of thin plates whose thickness is negligible compared with the overall dimension
Sideframe Structural member comprising side wall of vehicle, usually comprising:
- Sill Rail forming bottom member of sideframe (sole bar)
- Waist rail Rail forming centre member of sideframe at base of windows
- Roof rail Rail forming top member of sideframe (cant rail)
- A-post Pillar ahead of front door on sideframe
- B-C post Pillar between front and rear door on sideframe (centre pillar)
- D-post Pillar behind rear door on sideframe

Space framework Three-dimensional system of end-load carrying bars pinned together and normally built up from triangular bays
Specific strength (stiffness) Strength (stiffness) divided by the weight of the member or whole structure concerned
Spreader plate Subsidiary member that feeds in concentrated load over a wide area of the main member

Stabilizing member Subsidiary member attached to the main member and preventing buckling deflection so that full strength can be exploited

Stanchion Vertical member of portal frame

Stringer Intermediate end-load carrying member in stiffened panel

Swage Indentation of sheet to form-in flexural and endwise stiffness

Sway collapse Sideways motion of portal frame collapsing as a hinged mechanism

Symmetric bending Of a vehicle structure due to both wheels of an axle striking a bump simultaneously

Tension field beam Idealized beam in which the shear force in the web is considered to cause diagonal tension bands

Torque box System of plates forming a box-sectioned member which is rigid in torsion

Ultimate tensile stress (strength) Stress at the onset of rupture

Underbody (Understructure) Continuous base structure at wheel level

Web Central shear load carrying element of a beam

Wrinkling Instability failure of member, under end loading, into close pitch waves which are small relative to the length of the member

Yaw motion In the ground plane of a vehicle as it is steered or diverted from its path

Yield stress Stress at commencement of yield

CHAPTER 1

PRELIMINARY CONSIDERATIONS IN VEHICLE STRUCTURAL DESIGN

THE MORPHOLOGY OF VEHICLE STRUCTURAL DESIGN

Emancipation of the body designer

THE two giants of vehicle manufacture following World War I were Ford and General Motors in the USA. Initially Ford were predominant with their standardized ('any colour so long as it's black') Model T car and van but in the mid twenties General Motors gave birth to the 'stylist' we know to this day and the tables were turned on Ford.

Ford had fathered mass production and design standardization and other companies were quick to imitate; General Motors, however, also recognized another strong force in the market place, customer personal preference. They saw the sectorization of the market; they designed for particular market sectors and offered a proliferation of body shapes and colours. By the mid 1920s they were outselling Ford and the serious business of body design took its place among the Industry's professionalisms.

A stylist would 'package' occupants and luggage (or goods and passengers in a commercial vehicle) around a fairly standardized layout of engine and drive train. A phalanx of body design proposals would be presented to the company directorate and eventually a style with a projected appeal of about two years hence was visualized and chosen.

Although this pattern of events is still recognizable in most major quantity-car builders, improvements in technology have made a variety of mechanical configurations feasible and advancement in the science of ergonomics is gradually turning the traditional stylist into a 'conceptual designer'. This individual combines visual skills for aesthetic design with a more analytical approach to packaging. The conceptual designer has to deal with a wide variety of vehicles for special purpose as the market grows and becomes more sophisticated in its requirements. With a present-day knowledge of vehicle types it is hard to believe that a common basic layout was once adopted for truck, bus or family car, not so many years ago.

Expensive capital equipment used in forming vehicle bodies has limitations which also must be thoroughly understood by the conceiver of a new vehicle model; legal and safety constraints on design must be understood, too. The body draughtsman and detail designer must be aware of how the structural concept for a new vehicle affects the load paths through individual members to be designed and detailed.

Vehicle design is becoming more and more an exercise in packaging. Trucks, vans, buses, and cars all have to be designed within strict spatial envelopes. Power units, transmissions, suspensions, and support systems of all kinds have to be found a place among the occupants and goods. These have to be transported in comfort, safety, and often at very high speeds, and be allowed wide apertures for entering/dismounting or loading/unloading.

Too often nowadays it is a case of 'do we have enough room left for the structure?' rather than the structural requirements of the vehicle dictating the layout of the propulsion, transmission, and suspension systems, according to the optimum reaction of the load inputs involved. In the early days of horseless traction, of course, it was different. When one looked at a vehicle one would nearly always *see* a stout chassis frame, not just wonder whether or not there might be one somewhere within. But then, of course, the chassis, with its sub-systems, was manufactured in one plant and sent to another for its coachbuilt superstructure, so clearly the technique and the design method was different.

It was later on when economies of scale were realized in vehicle manufacture that design standardization of the chassis and all-metal construction of the body was to reverse the relative importance of the body and chassis groups. Prior to that, chassis design and development was considered separately from the body which in reality was a hardwood frame with pinned-on aluminium or steel panels.

For the benefit of the specialist builder it is advisable to remember the purpose of the structure, whether it be for fun car, family car, bus, or truck. The purpose of the car structure is to link up the mounting points for the vehicle's rear suspension and final drive, front suspension and steering, engine and gearbox, tank for fuel, and seats for occupants. It requires a rigidity to maintain accurate handling, lightness to reduce inertia and rolling drag, and toughness to sustain punishing fatigue loads from road, power unit, and driver.

Most trucks still retain chassis frames; in buses they are being 'developed out' and in cars they have virtually disappeared. Even in racing and sports cars unitary construction has become the rule rather than the exception and the need for an introduction into structural analysis of these integral structures has come about.

In this chapter we shall look at the basic scantlings of the structure and the purpose of the various elements, the way structures are developing, and consider the input loads which can be used for design. Then we shall examine the layout of the vehicle and see how it is affected by aerodynamics, packaging of the mechanics, and packaging of the occupants. As the book develops we shall study the behaviour of thin-walled beams in bending and torsion, the way in which the real-life car and bus structures are idealized for analysis and see how load paths through structures can be determined. Then we shall see how structures can be designed for impact and fatigue resistance and how they are affected by the manufacturing technique involved. Finally, we shall consider the special problems of truck and van structures both chassised and chassisless, with a taste of more advanced structural analysis techniques.

In their book *Racing and Sports Car Chassis Design* (Batsford,) Costin and Phipps examined that important intermediate stage in sports/racing car structural evolution, the space framework. They provided a useful analysis of the load paths in frameworks and a study of the rationale behind contemporary structures. Figure 1.1 shows four types of structure reproduced from their work; the twin tube Lister Jaguar at A, the multi-tube Cooper Monaco at B, the 'true' space frame of the Lotus Nineteen at C, and the chassis/body unit of the Lotus 25/33 at D. The authors pointed out that a special approach was necessary to the analysis of the final type of structure; it is hoped that this book will help to fulfil that need.

Elements of the structure

Many of the observations of Costin and Phipps on the initial approach to vehicle structural design are as valid for thin-walled tubular shell structures as they were for space frames of the 1950s and 1960s. Thus the initial approach to 'shell' structure design can commence with a drawing layout aimed at optimum positioning of the main structural elements. The specialist builder will need a minimum of four drawings to one fifth scale to depict front, side, and rear elevations, and a plan view. Since thin-walled shell construction involves large hollow closed members, additional cross sections might also be useful.

The first step is to draw in front and rear roll centre heights, from a suspension analysis, and the hub support members at either end. This enables the suspension linkage to be laid out and the vertical wheel travel determined. Next draw in front/rear axle centre lines, ground line, and undertray line to give adequate ground clearance at full bump. Reach diagrams for the drivers (see Chapter 2) and occupants will help to fix pedal and hand control positions from ergonomic theory, and the space left for packaging engine, transmission, radiator, and fuel tank can then be analysed.

Structure design starts with the major transverse members or bulkheads, generally one each at front and rear axles and two in between for fascia and seat back, positioned according to whether the car is front or rear

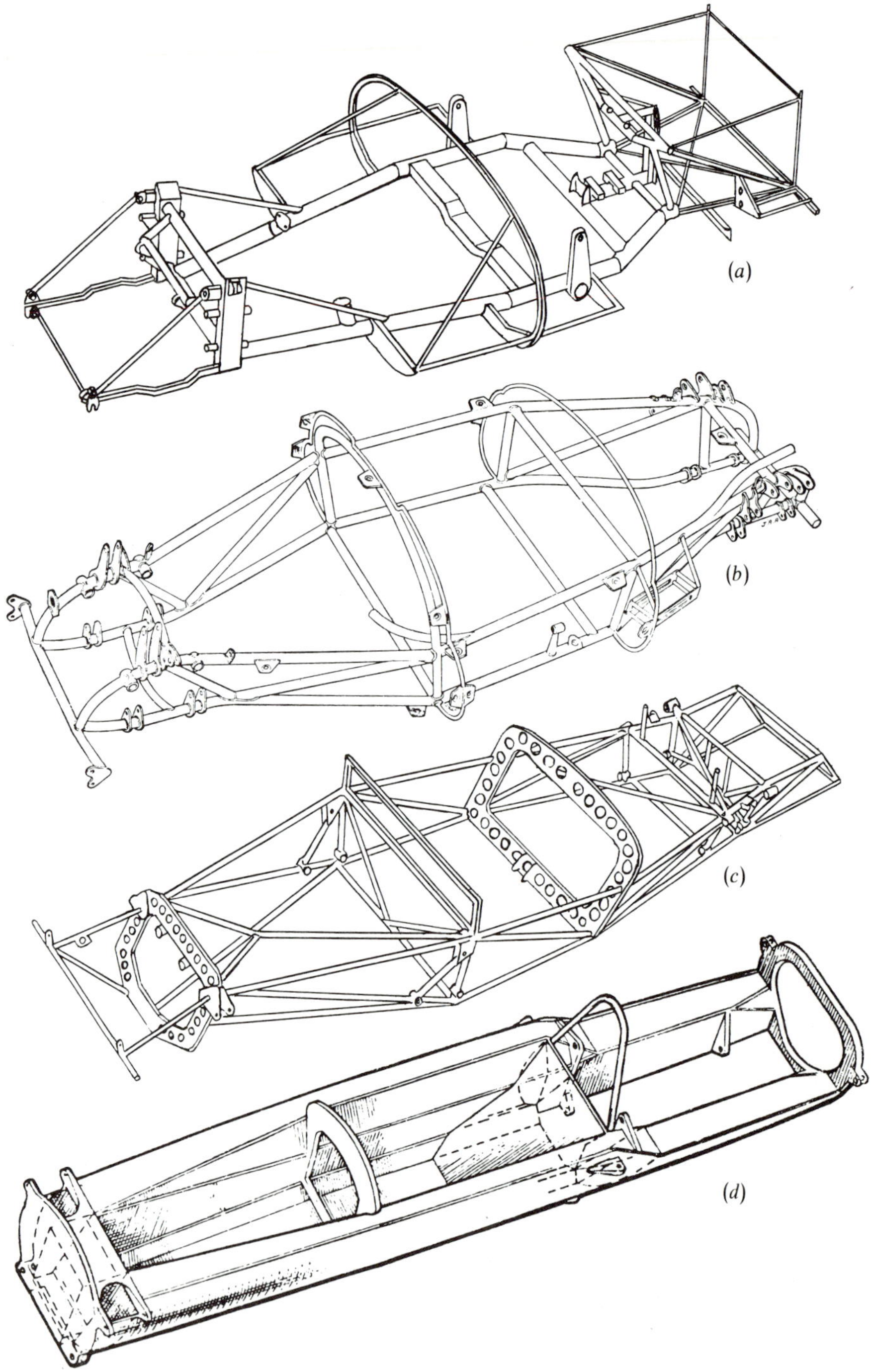

Fig. 1.1. Evolution, through space frame, of chassis frame to shell structure

engined. On a front-engined sports car the front bay is normally occupied by the power unit, the central bay by the occupants' legs, and the rear bay by the seats; on a rear-engined car, feet and legs are in the front bay, seats are in the centre, and rear bay is occupied by the engine/transmission.

Bulkheads through which driver's legs and engine/transmission protrude must have large cut-outs; these therefore need careful design to ensure that they are structurally stable. All the transverse bulkheads must be part of a structure which is rigid enough for minimum bending and torsional deflection. Large regular sectioned torque boxes are required in forming the longitudinals which must be deep enough for adequate bending rigidity. In the design shown in Fig. 1.2(a) for the Pininfarina Sigma racing car structure, the two rearmost bulkheads are joined by deep box beams, themselves cut out in the webs to allow for movement of the drive shafts. The front three bulkheads are joined by more widely spaced box beams in turn stabilized by outrigged half oval section members forming the fuel tanks and the streamline fairings for the wheels. In the critical area of the engine/transmission mountings additional members link the two support bulkheads, seen tapering in plan view. Figure 1.2(b) shows a contemporary road car understructure in the 'high performance' category.

The detail design of mountings on the bulkheads and longerons for engine/transmission, and all the others involving concentrated loads, is of crucial importance. Thin walled shell structure design is only successful if the full section proportions are exploited in resisting the loads, so that the design of web stiffeners, stabilizing rings, and spreaders need attention.

Deflection as a design criterion

Colin Campbell in his book *The Sports Car* (Chapman and Hall) nicely describes the limitations of the traditional chassis, likening it to a good bridge structure. He points out that whereas a good bridge is designed to sit on solid piers a good chassis has to sit on a road that flutters and wobbles underneath—like the waves of a choppy sea. When diagonally opposite wheels lift simultaneously on an uneven surface severe racking strains are imposed and a considerable twist can take place. Torsional deflections of the somewhat flexible coachbuilt bodies and disturbance of the steering geometry could result in a rough ride.

Further advances in vehicle development were to be an even greater threat to the traditional design of chassis frame. As car buyers, and later bus riders, demanded more comfort, softer suspensions had to be considered. These required much larger vertical wheel deflections than were possible with the beam-axled front (steering) suspensions of the early vehicles. Large vertical wheel motions with beam-axled suspensions would

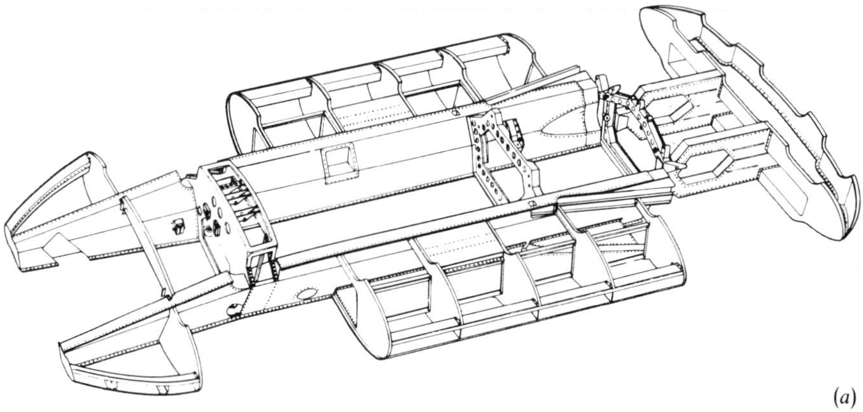

(*a*)

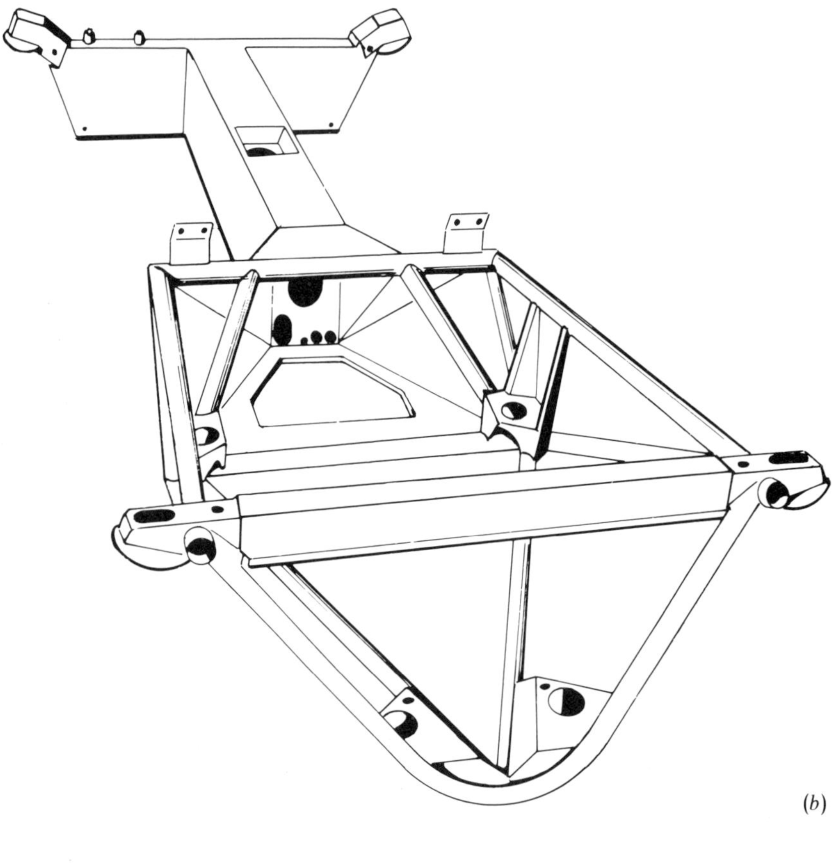

(*b*)

Fig. 1.2. (a) Pininfarnia Sigma racing car structure

have meant severe changes in wheel inclination and the resultant development of high gyroscopic torques. Only by providing independent suspension of the front (steered) wheels could the high wheel displacements be achieved, which, with minimal camber and steer angle change, were the prerequisites of a soft ride. But with the discarding of the beam axle and single piece track rod of the coupled front suspension a new rigidity was required of the vehicle structure to which the independent front suspension linkages were anchored.

Campbell considers the torsional stiffness of a chassis frame of substantial square tubular section shown in Fig. 1.3. When one wheel rides a road surface bump the suspension spring is compressed and the spring load is transmitted to and causes deflection of the frame. For spring stiffness c newtons per millimetre (N/mm) the deflection produced on the spring by the bump is d mm, and the load transmitted to the frame cd N. If the torsional stiffness of the frame is C N/mm measured vertically at the suspension anchorage points, the torsional deflection of the chassis $D = cd/C$ mm. For a spring rate of 100 lbf/in (15·7 N/mm) and a road wheel deflection of 4 in (102 mm), to limit the frame deflection to 0·03 in (0·76 mm), the torsional stiffness of the frame will need to be 12 800 lbf/in (2230 N/mm).

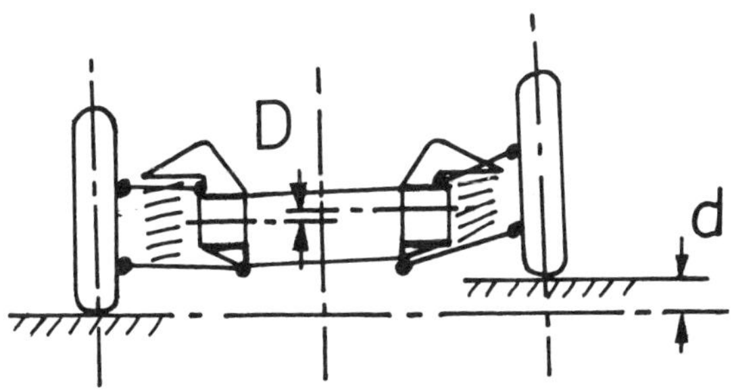

Fig. 1.3. Torsional deflection of a vehicle frame

It is more usual to express torsional stiffness in lbf ft/deg (Nm/deg) of twist between front and rear 'axles'. In a typical family saloon 4500 lbf ft/deg (6100 Nm/deg) should be a minimum with 5000–5500 (6500–7500 Nm/deg) preferred. It is thus apparent that criterion for vehicle structures is normally for stiffness rather than strength, the structural analyses being of deflections rather than stresses. Mid span bending deflections for a car should not exceed 0·05 in (1·27 mm) and door aperture deformation should not exceed 0·05 in (1·27 mm) for a 1500 lb (680 kg) mid span load.

Contemporary car structure classification

Integral: in which the main loads are carried by the members making up the structure. In practice this means that all the bodywork, except doors and hatches, is load carrying.
Semi-integral: in which there is a separate chassis frame but the remainder of the structure contributes substantially to the overall stiffness.
'*Punt*' *structures*: in which the floor is a box section with adequate torsional stiffness so that open-sports or lightweight GT bodywork can be fitted.

For the small family saloon Maurice Platt suggested, in his paper to the Institution of Mechanical Engineers: *The Structure of the Automobile* (1960 Crompton Lanchester lectures), that the ascendancy of the unitary (integral) body in Europe was due to the following.

1. More effective use of material resulting in weight saving or higher strength and rigidity for given weight.
2. Some saving in factory cost at the penalty of higher tooling cost.
3. 'All of a piece' construction giving freedom from destructive relative movements between chassis structure and body shell.
4. Well suited to the design of cars with low floors.

Platt points out that tests of torsional stiffness of complete shells do not merely provide absolute values of stiffness. By placing numerous dial gauges along the length of the structure, graphs can be obtained such as in Fig. 1.4 which would reveal any undesirable discontinuities in stiffness along the body.

The discontinuities would reveal possible failure points in road service under the action of cyclical racking strains as the vehicle negotiates uneven surfaces. Clearly the reinforcing at such points and repeating the stiffness test in search of negligible discontinuity would lead to a more durable structure.

The main objectives in structural design are to make the best use of material by arranging for each member to support as near as possible its maximum load potential and to make the structure direct and continuous by providing an unbroken path from point of application to point of reaction. For the specialist builder the advantages of mild pressing steel enjoyed by the mass producers are insignificant on a one-off special or short series build. By careful examination of the specific strength and specific stiffness of different materials an optimum choice can be made, governed as well by the fabrication facilities available. Figure 1.5 shows specific stiffness of various materials to illustrate the advantages of light alloys. In considering material in sheet form the flexural stiffness is a sounder criterion than the direct strain.

Typical metal thicknesses used in private car construction are 0·036 in

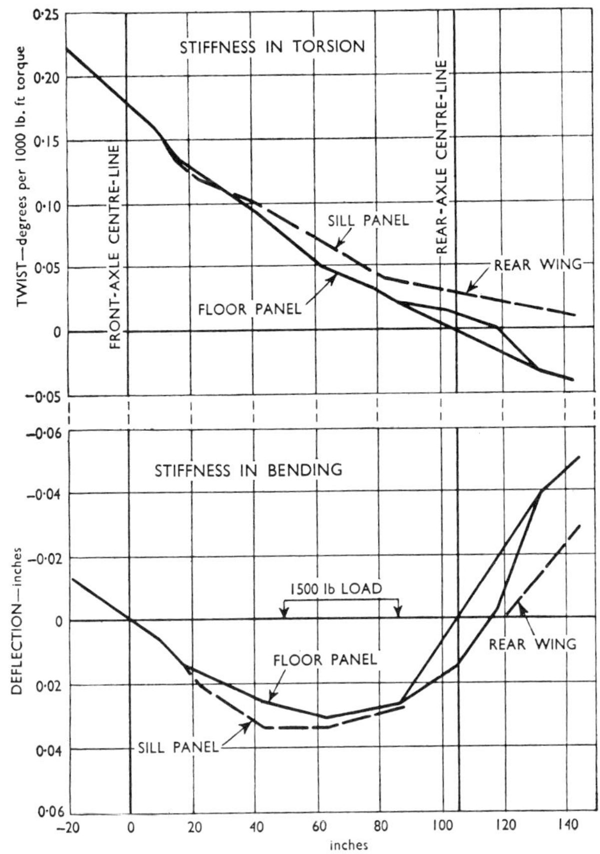

Fig. 1.4. Torsional and bending stiffness plots over the length of a car body

(0·915 mm) for large area panels such as roofs, quarters, doors, and floors; these can reduce down to between 0·030 and 0·026 in (0·765 and 0·66 mm) where curved contours add stiffness and 'tightness'. In major structural parts such as cross members, sills, rails, and pillars, thicknesses of 0·039 and 0·049 in (1–1·25 mm) are common with 0·064 in (1·625 mm) used for local reinforcements.

Panels are normally spot welded together, typical strengths for given panel thickness are shown in Table 1.1. Pitch for spot welds normally varies between 1 and $2\frac{1}{2}$ in (25·4 and 63·6 mm) according to strength and sealing requirements; minimum pitch is three times the spot-weld diameter. A spot-welded flange will normally be subject to twist because the shear load in the panel is unlikely to be applied through the centroid of the joint, Fig. 1.6. Each weld will be subject to load $Q \,.\, d_i^2/\sum d_i^2$ in the direction of loading together with a force $Q \,.\, L \,.\, r_i \,.\, d_i^2/\sum r_i \,.\, d_i^2$ (for spot-weld

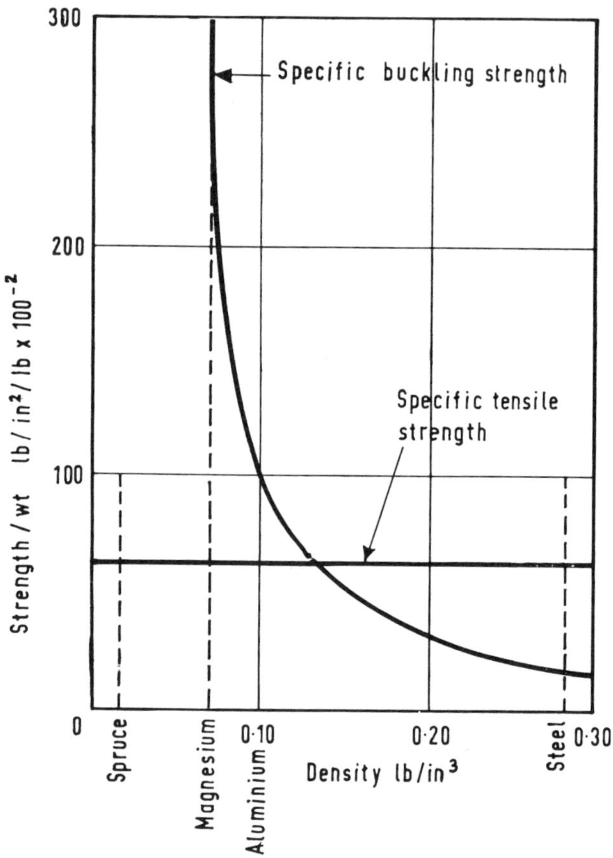

Fig. 1.5. Specific strength of body materials

diameter d_i) perpendicular to the line between the weld centre and the centroid, due to the torque of the applied load about the centroid.

In looking at structural configurations in integral structured cars Platt summarizes the main elements of the structure as an assembly of rails, sills, and pillars with practical sections as exemplified in Fig. 1.7. He sites the

TABLE 1.1

SWG	*Thickness (in) (mm)*	*Spot weld dia. (in) (mm)*	*Average shear failing load (ton) (tonne)*
20	0·036 (0·915)	0·18 (4·575)	0·63 (0·64)
18	0·048 (1·22)	0·22 (5·6)	0·80 (0·813)
16	0·064 (1·62)	0·25 (6·35)	0·88 (0·895)

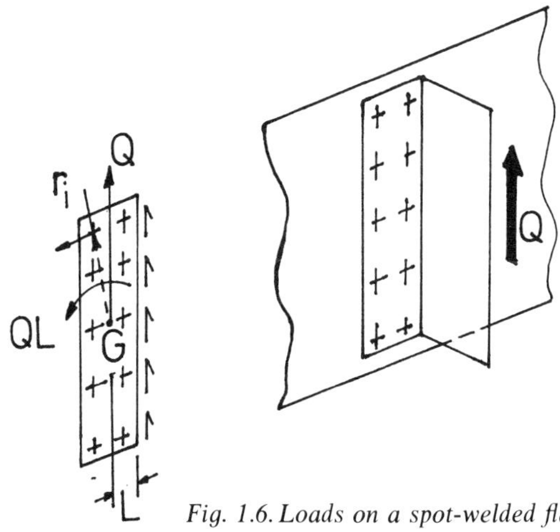

Fig. 1.6. Loads on a spot-welded flange

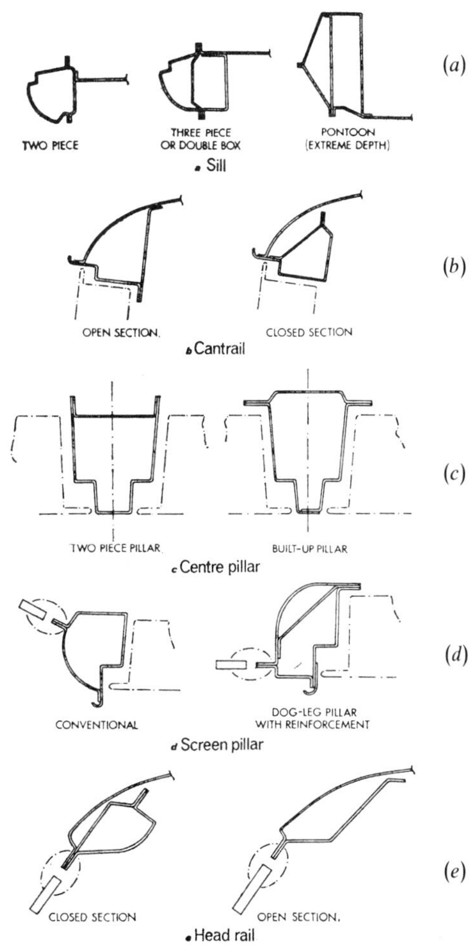

Fig. 1.7. Typical car body member sections

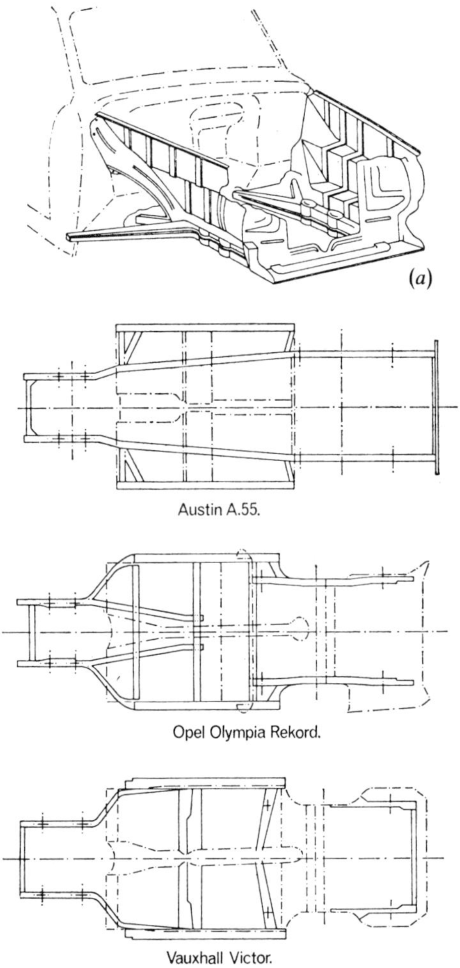

Fig. 1.8. (a) Holden car front end structure (b) Typical car underframe configurations

GM Holden, Fig. 1.8(a), to demonstrate the triangulation in elevation and plan designed to provide rigid connection between front suspension and bulkhead. This design had proved successful even on gruelling roads in the Australian outback.

The underbody is still a very substantial contributor to the main structural strength and Platt contrasts traditional approaches to underbody layout (Austin, Opel, Vauxhall) in Fig. 1.8(b), indicating the differences between 'continuous' and 'interrupted' underbody side members. In fact, more recent studies have shown that the various members which make up the total sideframe of the vehicle, not merely the sills, contribute considerably

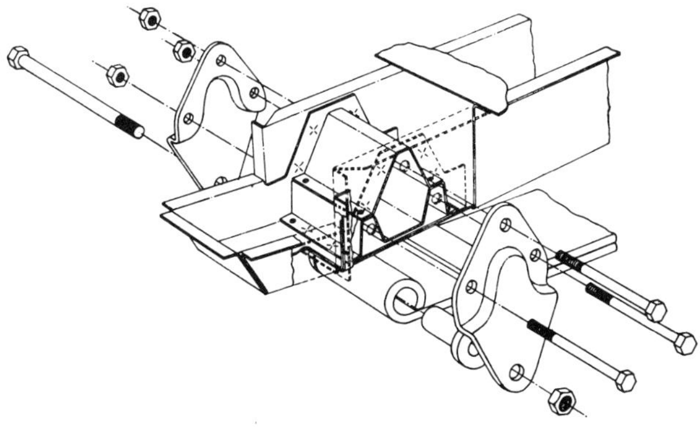

Fig. 1.9. Car rear spring front hanger attachment

to the overall beaming strength of the saloon car structure. Important contribution of the cross members is noted in the Opel design which like the Vauxhall of the period relies on sills for beam stiffness amidships. The Vauxhall relies on cranked longitudinals for front-end continuity and at the rear a special reinforced wheel arch incorporating a double walled rear wing. Structural continuity is more easily secured towards the back of the body than at the front end.

Attachment of mechanical units to an integral body needs special attention to the spreading of load in avoiding local distortion and failure, also in limiting undesirable relative movement. The rear spring front hanger bracket is an important area in conventional leaf sprung front-engine/rear-drive layouts. This mounting has to transmit braking and drive loads horizontally as well as vertical loads due to weight and braking/driving torque reaction, and side loads due to cornering. The example illustrated in Fig. 1.9 shows that the boxed (longitudinal) body member has an internal stiffener and side plates with through bolts.

THE DYNAMIC DESIGN LOADS ON A ROAD VEHICLE

Any worthwhile structural analysis clearly requires an understanding of the pattern of loads applying to the vehicle. Kenneth Garrett writing in *Automobile Engineer*, February 1953 ('Automobile Dynamic Loads'), nicely describes the combination of applied loads arising in acceleration, braking, cornering, and kerbing by considering four design cases representing the worst combinations of loadings which could affect the vehicle structure at any instant.

Dynamic measurements on test vehicles have shown peak accelerations of $3g$ recorded in the vertical direction. In the horizontal plane transverse cornering forces and longitudinal braking forces are limited by the adhesion of the tyre to the road and a limiting retardation figure of $1g$ has become acceptable. Garrett suggests multiplying these accelerations (or decelerations) by 1·5 as a factor of safety in design to arrive at the corresponding maximum loads. Thus maximum accelerations are

$\pm 4{\cdot}5g$: vertical (hitting a bump)
$\pm 1{\cdot}5g$: fore-and-aft (braking and accelerating)
$\pm 1{\cdot}5g$: cornering LH or RH

The situation of striking a kerb head-on is examined in Fig. 1.10. If the vehicle brakes are 'off' the resultant force can only act through the axis of

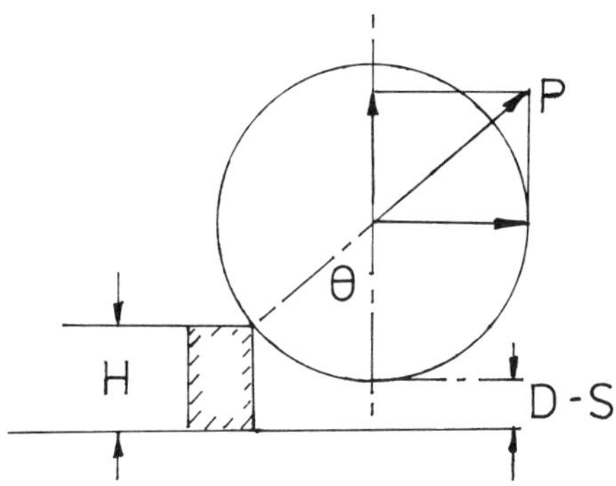

Fig. 1.10. Loads due to striking kerb

the rotating wheel. For static vertical load reaction per wheel R giving a vertical force at bump of $4{\cdot}5R$ then the resultant load along a line through the wheel axis and the centre of contact of the bump is $P = 4{\cdot}5R/\cos\theta$ which has a horizontal component

$$\frac{4{\cdot}5R\sin\theta}{\cos\theta} = 4{\cdot}5R\tan\theta$$

If the motion to the full bump travel D of the suspension has been very rapid so that the car body has only risen a negligible distance, the effective height of the kerb is $H - (D - S)$ where S is the static deflection (under the weight of the vehicle) of the suspension. Kerb height H is normally taken as 6 in (152·4 mm) (allowing for tyre deflection).

Let us examine the loads due to cornering in Fig. 1.11 in terms of front and rear static wheel reactions R_A and R_B. When the vehicle is almost overturning the laden weight W acts through the outer wheels only; the limiting centrifugal force adds up to the vector sum of the radial ground forces $\mu(R_A + R_B) = \mu W$ for friction co-efficient μ. So that in line with and perpendicular to the longitudinal axis of the vehicle

$$\mu(R_A\cos\phi + R_B) = \mu W\sin\psi$$

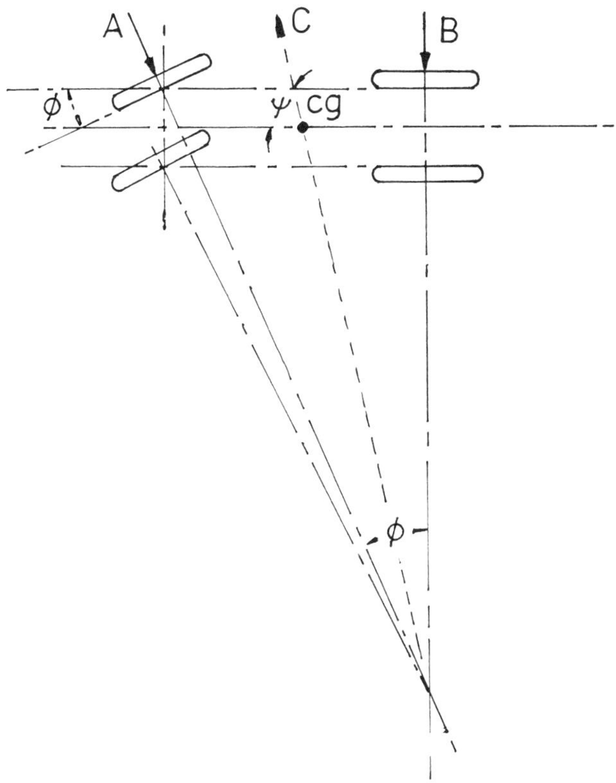

Fig. 1.11. Loads due to cornering

and

$$\mu R_A \sin \phi = \mu W \cos \psi$$

whence

$$R_A = W \cos \psi / \sin \phi \qquad \text{and} \qquad R_B = -W \cos (\psi + \phi) / \sin \phi$$

ϕ and ψ are obtained from the layout of the vehicle.

He now considers the four design load cases:

Case 1: Hitting a bump while braking as might occur if a driver sees an unexpected obstacle or pothole, applies his brakes, and hits it. There will be

Vertical load due to bump $= R \times 4{\cdot}5$
Rearward load due to bump $= R \times 4{\cdot}5 \times \tan \theta$
Load transfer due to bump $= R \times 4{\cdot}5 \tan \theta \, . \, \bar{y}/2B$
Load transfer due to braking $= 1{\cdot}5 W\bar{y}/2B$
Rearward load due to bump $= 1{\cdot}5R + 1{\cdot}5W\bar{y}/2B$

with wheelbase B, track T, height of centre of gravity above ground $\bar{y}$, and single wheel reaction R. These add up to a

$$\text{Vertical component} = 1{\cdot}5\left\{3R\left(1 + \frac{\bar{y}}{2B}\tan\theta\right) + \frac{W\bar{y}}{2B}\right\}$$

$$\text{Rearward component} = 1{\cdot}5\left\{R(3\tan\theta + 1) + \frac{W\bar{y}}{2B}\right\}$$

$$\text{Lateral component} = 0$$

and similar additions may be made for the three other cases:

Case 2: Cornering

Case 3: Hitting a bump while accelerating

Case 4: Hitting a bump whilst cornering

Case 2: $$\text{Vertical} = 1{\cdot}5W\left(\frac{\cos\psi}{\sin\phi} + \mu\cos\psi\frac{\bar{y}}{2B}\right)$$

$$\text{Rearward} = 1{\cdot}5\mu W\cos\psi$$

$$\text{Lateral} = 1{\cdot}5\mu W\frac{\cos\psi}{\tan\phi}$$

Case 3: $$\text{Vertical} = 1{\cdot}5\left\{3R\left(1 - \frac{\bar{y}}{2B}\tan\theta\right) + \frac{W\bar{y}}{2B}\right\}$$

$$\text{Rearward} = 4{\cdot}5R\tan\theta - 0{\cdot}75W$$

$$\text{Lateral} = 0$$

Case 4: $$\text{Vertical} = 1{\cdot}5\left\{R\left(3 + \frac{2\bar{y}}{T}\right) - \frac{\bar{y}}{2B}(\mu W\cos\psi + 3R\tan\theta)\right\}$$

$$\text{Rearward} = 4{\cdot}5R\tan\theta$$

$$\text{Lateral} = 1{\cdot}5\mu\frac{W\cos(\psi + \phi)}{\sin\phi}$$

Examples of loading cases

Kenneth Garrett writing later in *Automobile Engineer*, March 1953 ('Structural Design') provides examples of loading cases 3 and 4 in considering the rear end structure of a small car. The following data were substituted in the formulae: $R = 610$ lbf (2710 N), $\theta = 48°$, $W = 2060$ lbf (9170 N), $\bar{y} = 27$ in (0·687 m), and $B = 77$ in (1·956 m) then $T = 43$ in (1·094 m), $\psi = 78°$, $\phi = 24°$, and $\mu = 1{\cdot}0$ so that case 3 loads are: 2940 lbf

(13·08 kN) vertical, 495 lbf (2200 N) rearward, and zero lateral. Then case 4 loads are: 3410 lbf (15·15 kN) vertical, 2040 lbf (9080 N) rearward, and 1570 lbf (6980 N) lateral—the worst case.

The rear spring front hanger shown in Fig. 1.12 is attached to the heel board and rear floor well of the car structure; the two brackets are spot

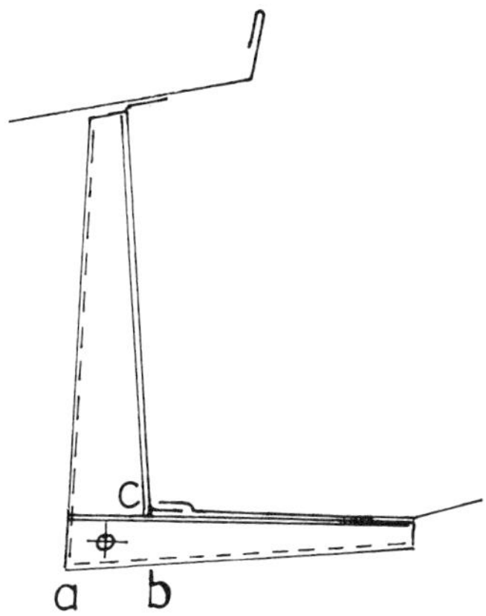

Fig. 1.12. Rear spring front hanger attachment

welded where they overlap. Considering the case 4 loading vertical component, and given that the spring shackle angle is 27 degrees from the vertical at full bump, the rearward directed component is

$$3410/2 \times 1/\tan (90 - 27) = 870 \text{ lbf } (3862 \text{ N})$$

This must be added to the drag load of 2040 lbf (9080 N) for a total of 2910 lbf (1·295 kN). Assuming spot-weld shear load capacity is 275 lbf (1220 N) the horizontal load must be transferred to the floor by a minimum of 2910/275 = 11 spot welds. This number is doubled to allow for dynamic effects. The load is tensile on the bracket so that sectional length required of 18 swg sheet steel is

$$2910/(16 \times 2240 \times 0{\cdot}048) = 1{\cdot}64 \text{ in } (41{\cdot}7 \text{ mm})$$

for a stress limit of 16tonf/in^2 (24.4 N/mm^2).

But space requirements at the spring eye dictate a 1·75 in (44·5 mm) wide × 1 in (25·4 mm) deep top hat section with 0·625 in (15·86 mm) flanges giving a section length of 5 in (127 mm) which reduces the maximum tensile stress to 5·25 tonf/in^2 (8 N/mm^2). Owing to the 1570 lbf (6980 N) side load, bending stress will be superimposed. 785 lbf (3490 N) is assumed to apply to each spring eye. When two lugs of a channel share side load it is usual to assume 2/3 of the load on each lug, in this case 524 lbf (2321 N). The length abc in Fig. 1.12 is 3 in (76·3 mm) and the depth 2 × 0·048 = 0·096 in (2·44 mm). Thus (where b and d are breadth and depth of section) section modulus $Z = bd^2/6 = 0{\cdot}046\ \text{in}^2$ (0·2965 cm²). With a mean moment arm of 2 in (50·8 mm), bending stress $= M/Z = 524 \times 2/0{\cdot}046 \times 2240 = 10{\cdot}2\ \text{tonf/in}^2$ (15·3 N/mm²) for bending moment M.

Testing the ability of the heel board to withstand the vertical loading through the spring pin hanger bracket is carried out by considering forces in the plane of the panel, Fig. 1.13. The weak point at section GH is shown

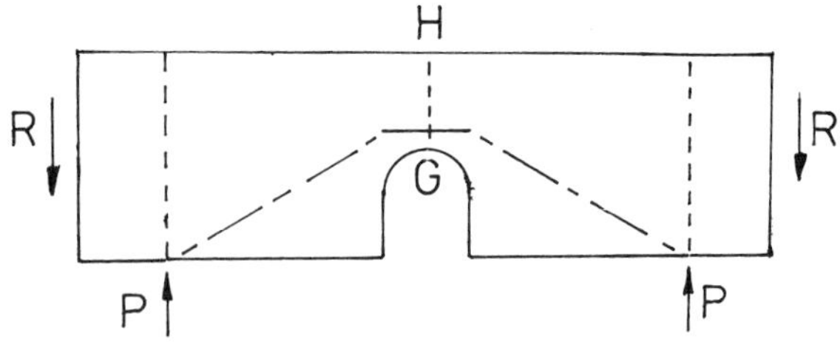

Fig. 1.13. Diagrammatic view of car heel board

in Fig. 1.14, the necked section over the propeller shaft tunnel. Loads P and reactions R, on the vehicle sideframe put the heel board in pure bending. With 5 in (127 mm) inset of the spring hangers from the sideframe, bending moment at GH is 5 × 2940/2 = 7350 lbf in (832 Nm).

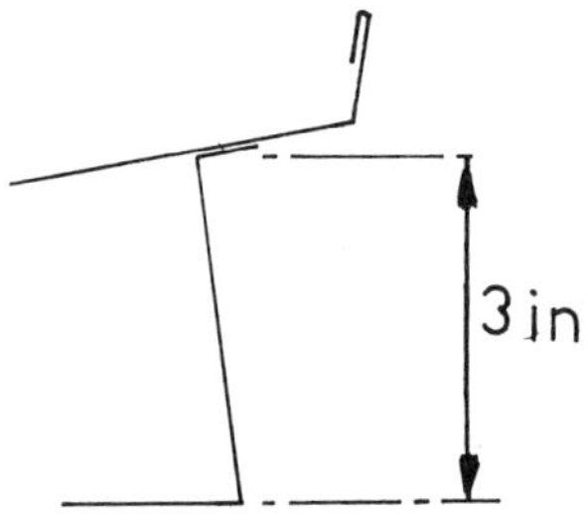

Fig. 1.14. Section GH through heel board in Fig. 1.13

Distance between the neutral axis of the flanges on the section GH is 2·75 in (70 mm) so that flange end load = 7350/2·75 = 2680 lbf (11·9 kN). For a limiting compressive stress of 12 tonf/in² (18·35 kN/cm²), sectional area of metal required = 2680/2240 × 12 = 0·1 in² (0·645 cm²) which is equivalent to 2·8 in (71·2 mm) of 20 swg steel. This is in fact accounted for by adding in 16 × metal thickness of adjacent panels (i.e., turned-up flange of seat pan, and section of seat pan either side of the heel board intersection).

Next loading cases 1 and 2 are considered by Garrett at the front of the car. Here $R = 3904/2 \times 4{\cdot}5 = 434$ lbf (1930 N) and case 1 loads become 2750 lbf (12·24 kN) vertical, 2640 lbf (1·75 kN) rearward and zero lateral. Case 2 loads are 1690 lb (7520 N) vertical, 642 lb (2855 N) rearward and 1440 lb (6410 N) lateral. Also the braking torque is considered assuming $1g$ retardation ($\mu = 1{\cdot}0$). With vertical reaction per wheel

$$4{\cdot}5 + W\bar{y}/2B = 797 \text{ lbf (3540 N)/wheel} = \text{drag load}$$

So with tyre rolling radius of 11·25 in (286 mm), brake torque = 797 × 11·25 = 8950 lbf in (1010 Nm).

For case 1, the loads to the four wishbone bearings of the independent front suspension shown in Fig. 1.15 are next investigated. In the figure

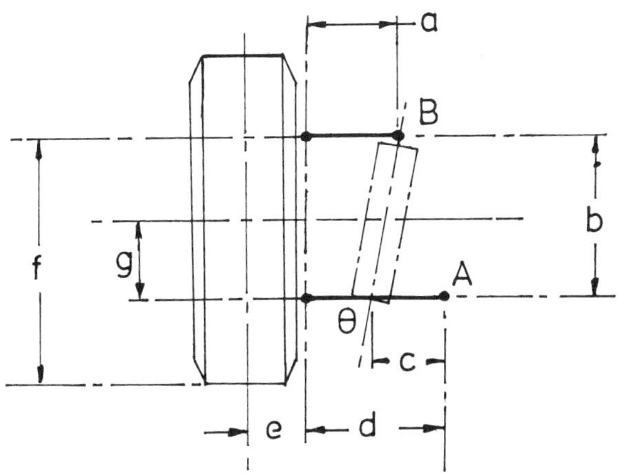

Fig. 1.15. Independent front suspension geometry

$a = 7$ in (178 mm), $b = 7{\cdot}5$ in (190 mm), $c = 5{\cdot}5$ in (140 mm), $d = 10{\cdot}75$ in (273 mm), $e = 3$ in (76 mm), $f = 15$ in (381 mm), $g = 3{\cdot}75$ in (95 mm), and $\theta = 80°$. Moments about A give vertical component of spring and bump stop reaction:

$$2750 \times 10{\cdot}75/5{\cdot}5 = 5370 \text{ lbf (23·8 kN)}$$

and a horizontal component

$$(5370/\sin 80°) \times \cos 80° = 946 \text{ lbf (4200 N)}$$

reacted at A. Then moments about the outer bearing, for inner bearing vertical reaction

$$(5370 \times 5{\cdot}525)/10{\cdot}75 = 2620 \text{ lbf (11·65 kN)/pair}$$

Differential horizontal loads at outer bearings due to offset of wheel centre = 2750 × 3/7·5 = 1100 lbf (4890 N).

Drag loads are assumed to apply either with wheel at full bump when upper wishbone bearing will take most, or at normal laden position in which a large proportion will be reacted by the lower wishbone. At full bump, top bearings will take (6·5/7·5) × 2640 = 2480 lbf (11·01 kN)/pair and differential load due to wheel offset = (2480 × 10)/5 = 4960 lbf (22 kN). At normal ride height lower bearings take 3·75/7·5 × 2640 = 1320 lbf (5870 N)/pair and the differential load = 1320 × 13·75/8·5 = 2140 lbf (95·10 kN). Differential load due to brake torque = 8950/7·5 = 1200 lbf (5330 N) on outer bearings. Differential load to upper inner bearings = 1200 × 10/5 = 2400 lbf (10660 N). Differential load lower inner bearings = 1200 × 13·75/8·5 = 1940 lbf (8590 N).

The loads are then summed for each inner bearing for loading case 1 and 2. These are shown in Figs 1.16 and 1.17 for which

	a	*b*	*c*	*d*	*e*	*f*	*g*	*h*	*i*	*j*	*k*	*l*
Case 1	0	3110	1240	0	2010	1240	3047	1400	660	5103	1400	660
Case 2	0	774	289	0	1538	289	1330	805	160	292	805	160

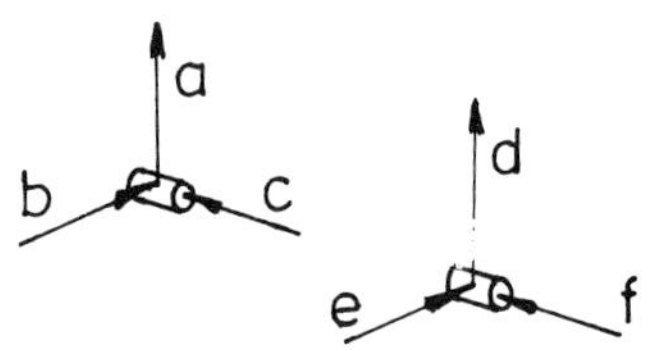

Fig. 1.16. Loads on inner bearings of i.f.s.

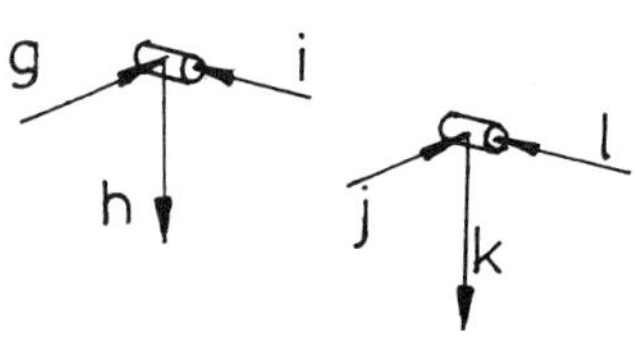

Fig. 1.17. Loads on inner bearings of i.f.s.

The upper wishbone is integral with a lever-type suspension damper which is bolted to a platform supported by a vertical channel member, Fig. 1.18. For the shear attachment of the platform to the channel, case 1 is critical, the direct lateral shear being 3110 − 2010 = 1100 lbf (4900 N). A couple about the vertical axis through the centre of the spindle due to differential loads at the bearings is 3110 × 2·5 + 2010 × 2·5 = 12 790 lbf in (1444 Nm), which must be reacted in shear in the spot welds to the sides of the vertical

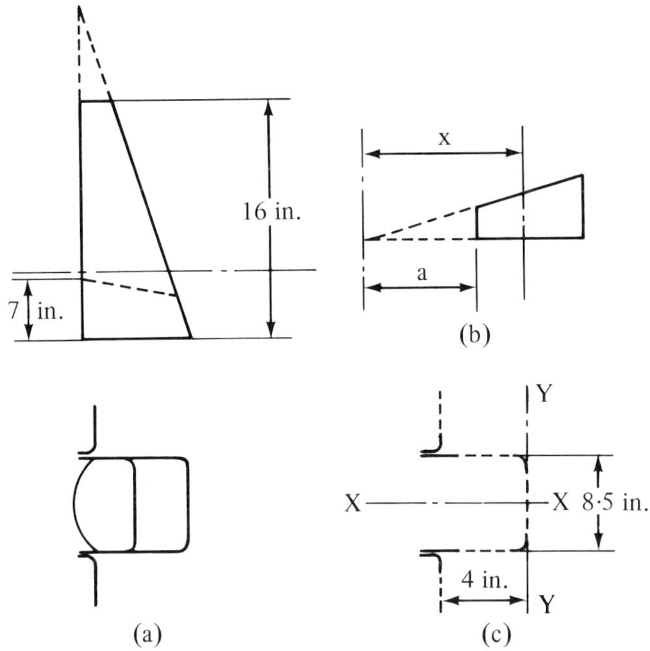

Fig. 1.18. Car front suspension attachment

channel. These sides are 8·5 in (216 mm) apart so shear per side is 12 790/8·5 = 1500 lbf (6660 N) and direct shear 1100/2 + 1500 = 2050 lbf (9120 N). Superimposed on this is the effect of the drag load of 2480 lbf (11·03 kN) acting along a line 8·5 in (216 mm) from the inner wall of the 18 swg channel. The direct shear will have to be taken by 2480/375 = 7 spot welds and the moment due to the offset must be taken by the spot welds on the side of the channel. This load of 2480 × 8·5/8·5 added to the 2050 lbf (9120 N) gives a total of 4530 lbf (20·15 kN) and at least 13 spot welds are needed.

CHAPTER 2

VEHICLE LAYOUT—CONSTRAINTS ON STRUCTURAL CONFIGURATION

PACKAGE design is a term used in the vehicle industry to describe the work a conceptual designer has to do in fitting the vehicle to its occupants on the inside, to its mechanical systems and to the exterior design envelope fixed by the stylist, aerodynamicist, legislator, or whomsoever. Vehicle layout is about package design and the following discussion will help the structural designer or draughtsman to appreciate the constraints on structural design.

We shall first discuss aerodynamic design, then examine by example a variety of mechanical layouts which have been used in recent years and see how they affect the structure. Then we shall look at ergonomic design and, finally, the layout of a complete vehicle in case study.

VEHICLE AERODYNAMICS—AIR RESISTANCE AND STABILITY

Increases in vehicle speed in recent decades have increased the prominence of air resistance among the various other forces, such as rolling and inertia resistance, which resist the motion of a vehicle. An aerodynamic law is established which relates air resistance r_a to vehicle speed (relative to the air) V_r, air density ρ, projected frontal area normal to the direction of travel A, in

$$r_a = \frac{c_a \cdot \rho \cdot A \cdot V_r^2}{2g}$$

where g is gravity acceleration and c_a is the drag coefficient which has unique value for a family of geometrically similar shapes. Typical values for c_a are 0·4–0·5 for a private car, 0·25–0·3 for a racing car, 0·6–0·7 for a bus and 0·8–1·0 for a truck. For comparison r_a value for a sphere is 0·47 and for a streamlined body is 0·1.

Air density varies with atmospheric temperature t, and pressure p, according to the law $\rho = 144\ p/R\ .\ t$ where R is the gas constant = 53·3 for air. For standard atmospheric conditions $\rho = 0{\cdot}0763$ lbf/ft^3. Physical origin of air resistance on vehicles is derived from three sources: (a) Drag resistance which is a function of aerodynamic shape; it is affected by protruding objects like mirrors. The value of r_a is particularly affected by body rear end shape which determines the amount of turbulence in the wake of the vehicle. (b) Skin friction of the body for typical surface finishes accounts for about 10 per cent of the air resistance. (c) Air flow through the vehicle for engine cooling and interior ventilation purposes.

The product of the drag coefficient and frontal area is the key variable from vehicle to vehicle. This product can actually decrease with a *rise* in frontal area if a protruding part is covered, say, with a streamlined fairing. Typical values of A in square feet (square metres) for frontal area are: small cars 14–20, (1·3–1·85 m^2); large cars 22–30, (2·04–2·70 m^2); buses and trucks 60–80 (5·6–7·4 m^2). Areas can be estimated as $0{\cdot}9 \times$ body height $\times$ wheel track.

In terms of engine power required to overcome air resistance this is equal to $0{\cdot}0026\ c_a \,.\, A \,.\, V_r^2 \,.\, V/375$ for standard air density, where V is the geared road speed. Thus in still air conditions, power required to overcome air resistance relates to the cube of vehicle speed, and the importance of aerodynamic drag at higher speeds becomes obvious.

In his article 'The aerodynamics of the modern car', Frank Costin, writing in *Automotive Engineer* (October–December 1975) showed in graphical form, Fig. 2.1, the effects of changes in drag coefficient on power

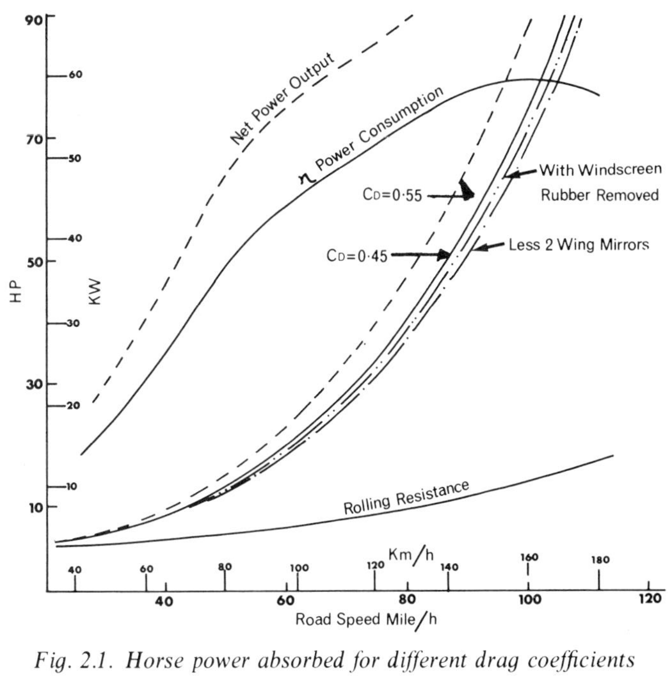

Fig. 2.1. Horse power absorbed for different drag coefficients

requirement. The plot of tyre rolling drag is comparatively small; note also the significant change in power requirement in changing from 0·45 to 0·55 drag coefficient. Even the effect of removing a rubber windscreen surround, standing proud by 0·5 in, (12·7 mm), and taking off two wing mirrors measuring 5 × 3 in (127 × 76 mm), is quite significant.

Areas of conventional car bodies considered by Costin to offer considerable scope for drag reduction are the 'sunken-rectangle' type radiator grille, the flight-deck type bonnet top, some designs of which actually cause reverse air flow, windscreen layout, particularly surround mouldings, rain gutters, and the rear of the 'top block', which normally is only a location for the rear window.

The important area of radiator intake requires an appreciation of the air flow characteristics at the radiator. Flow through it is at relatively low speed and during the cooling process there is a change in pressure. The desired condition is high pressure and low velocity at the front face of the matrix and a low pressure at the rear face. Also the air should be speeded up from that at the intake so that it can be returned to the outside air at the same or a higher speed.

For the design of undertrays it is recommended to allow as great a distance between the undertray and the road as possible, the shortest possible transverse span of flat surface, the largest practicable side radii, and a slight nose down incidence of the entire undertray. Figure 2.2 shows the profile of a car assumed to have no front air intake. As the air approaches the body it has to divide to flow around it; this point of diversion is termed the stagnation point where the 'dynamic pressure head' is $\frac{1}{2}\rho V_r^2$. As the air flows around the body a pressure distribution is developed as indicated. As the air velocity is increased over the roof a pressure drop occurs to cause a suction zone. Conversely, at the base of the windscreen the streamlines separate and a pressure results from the low velocity. Starting again at the stagnation point, as the air moves around the curved nose it speeds up and the pressure falls until at B atmospheric pressure is reached. The drag force is of course the summation of the horizontal components of the arrows around the pressure distribution. The contribution diminishes very rapidly as the distance from the stagnation point increases and therefore the nose shape is critical to the drag; surrounding lips which trap air should especially be avoided and careful rounding of flat surfaces such as headlamp mounts should be employed.

Drag reduction can also be obtained by reducing the 'wake' size, that is, maintaining unseparated air flow much further back on the body. Whereas the flow around a smooth aerofoil section, say, is streamline from nose to tail, a bluff car body inevitably causes a break up of the flow into turbulence. While flow remains streamline, air adheres to the car over a thin boundary layer whereafter it is sheared until the air velocity builds up

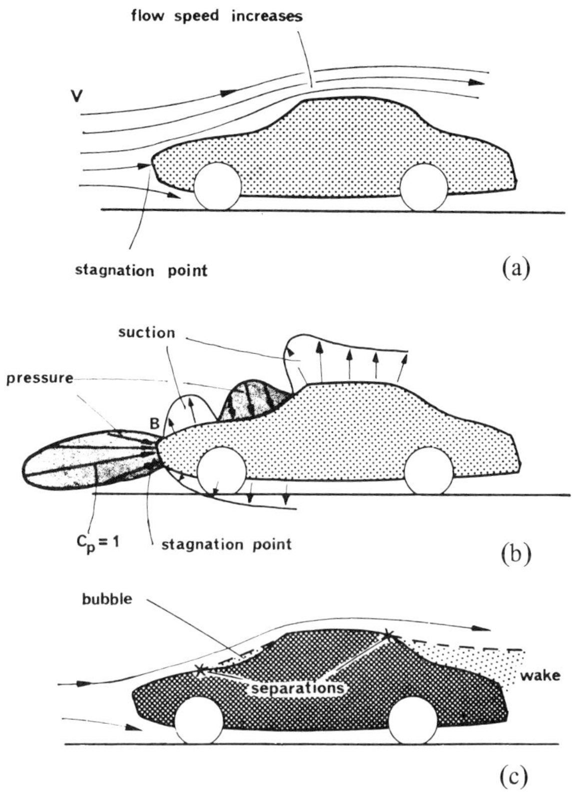

Fig. 2.2. Air flow and pressure distribution around a typical saloon car

to that of the main stream. Tangential viscous friction forces arise which add to air resistance whenever smooth flow is distorted by surface roughness or protuberances. Severe discontinuities in the surface profile can cause a separation of flow which is the commencing point for the turbulent wake. When flow reaches the corner of a windscreen, for example, it must slow down causing the pressure to rise, thus producing a separation. In this particular case flow would probably re-attach itself further down the vehicle and a bubble would form, Fig. 2.2(c).

Means should be provided for maintaining boundary layer as far back along the body as possible. Very gentle curvature can maintain attached flow right up to the rear of the vehicle. It is important to remember that the car is a three-dimensional shape and there is a flow around the sides as well as over the top. If the side windows on a saloon car are sloped well inwards not only will the blending in of the windscreen be easier but air will make its way around the sides to sweep the separation point further towards the tail of the car.

Writing in *M.I.R.A. Report* 1967/9 ('A rating method for assessing vehicle drag coefficients'), R. G. S. White provides a method of adding up the effects of the shapes of different areas of a car body to pre-estimate drag coefficient, in the blanked radiator condition, at the design stage. He considers the following areas (see Fig. 2.3): front end, plan outline (1); side

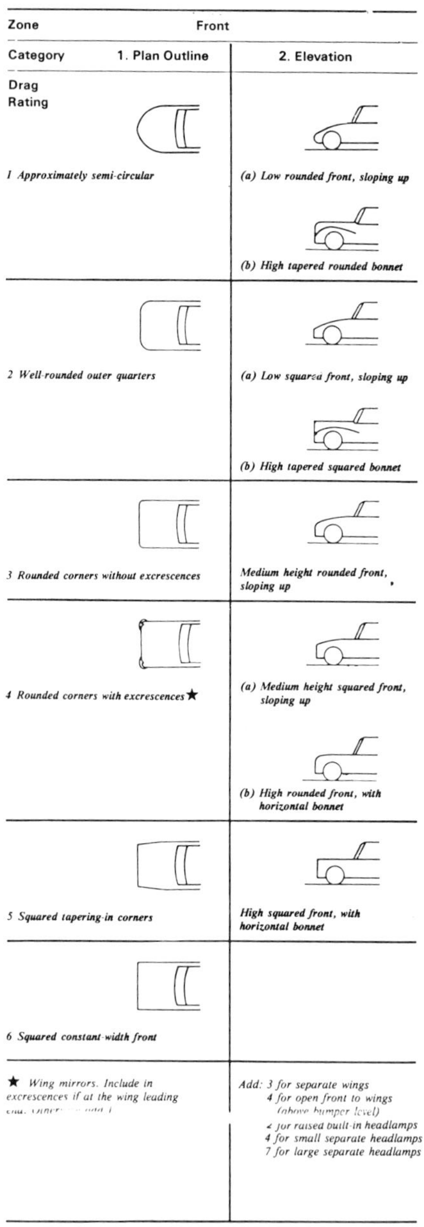

Zone	Windscreen/Canopy junction	
Category	3. Scuttle and wing cross-section	4. Windscreen plan
Drag Rating	 *1 Flush bonnet and wings, well-rounded body sides*	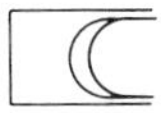 *Full wrap-round (approximately semi-circular)*
	 *2 High scuttle, low wings*	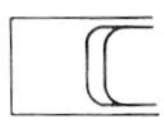 *Wrapped-round ends*
	 *3 (a) Bonnet flush with rounded-top wings*  *(b) High scuttle, with rounded-top wings*	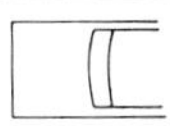 *Bowed*
	 *4 Bonnet flush with squared-edge wings*	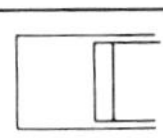 *Flat*
	 *5 Depressed bonnet, with high squared edged wings*	
		Add: 1 for upright windscreen 1 for prominent flanges or rain gutters

Zone	Rear canopy/boot	Lower rear quarters	Underbody
Category	7	8	9
Drag Rating	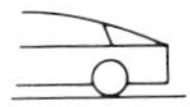 *1 Fast-back, i.e., roof line continuous to tail*	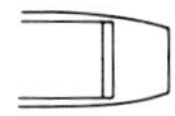 *Well-or-medium-tapered to rear*	*Integral, flush floor, little projecting mechanism*
	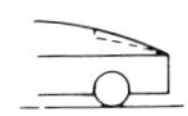 *2 Semi-fast back i.e., with discontinuity in line to tail*	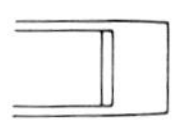 *Small taper to rear or constant width*	*Intermediate*
	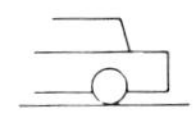 *3 Squared canopy with boot rear edge squared*	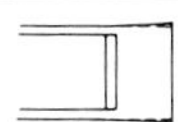 *Outward taper (or flared-out fins)*	*Integral, projecting structure and mechanism*
	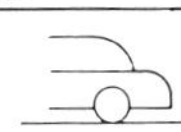 *4 (a) Rounded canopy with rounded boot* 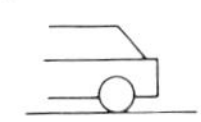 *(b) Squared canopy with short or no boot*		*Intermediate*
	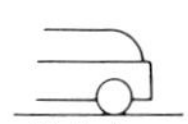 *5 Rounded canopy with short or no boot*		*Deep chassis*
	Add: 3 for high fins or sharp longitudinal edges to boot 2 for separate wings Note. In all the ratings in this column, the boot is assumed to be rounded laterally	*Note. Throughout table, the word "taper" or "tapered" refers to the plan view.*	

Zone	Canopy	
Category	5. Windscreen peak	6. Canopy plan
Drag Rating	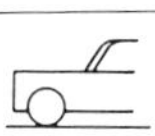 *1 Rounded*	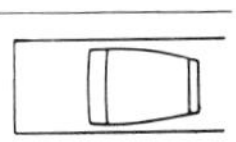 *Well- or medium-tapered to rear*
	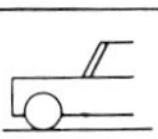 *2 Squared (including flanges or gutters)*	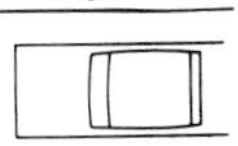 *Tapering to front and rear (max. width at BC post) or approximately constant width*
	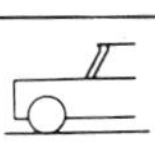 *3 Forward-projecting peak*	 *Tapering to front (max. width at rear)*

Fig. 2.3. Contributions to air resistance of different areas of a car body

elevation (2); windscreen canopy junction, scuttle/wing cross-section (3); windscreen plan (4); canopy zone, windscreen peak (5); canopy plan (6); rear canopy/boot (7); lower rear quarters (8); and underbody (9). Ratings shown for each configuration given in Fig. 2.3 are summed to a total rating and the drag coefficient obtained from; $c_a = 0{\cdot}16 + 0{\cdot}0095 \times$ drag rating. The author claims an accuracy of ± 7 per cent.

Aerodynamic stability

Stability in roll, pitch, and yaw needs to be considered. The first can be achieved by ensuring that centre of pressure (point of application of wind side force) is close to the roll axis of the car. Small deviations from the ideal can be tolerated if the vehicle does not have pronounced roll-steer characteristics. Distribution of lift forces acting on a vehicle determines the magnitude of the pitching moment which tends to vary the wheel loading with speed. A tail-end spoiler used to form the top edge of a drag-reducing cut-off tail also causes negative lift on the rear wheels and can be useful in increasing lift stability. Lightweight cars with centre of gravity well aft are less easy to control and are susceptible to zero or small negative lift at the front. Techniques such as drooping the nose can be helpful, which is equivalent to reducing the angle of incidence of an aerofoil.

In wedge-shape cars the top surface of the body provides a negative angle of incidence which results in the aerodynamic lift force being directed downwards. Also the under side of the body together with the ground forms a kind of venturi tube which in turn increases the suction effect to produce more negative lift. But for a very low slung car considerable turbulence develops with the complex interference of the boundary layers with reduced mean flow velocity and increase in pressure causing an overall lift; hence the use of negative lift devices on racing cars.

Yawing plane stability is probably the most important to the average motorist since it will govern the amount of steering correction required when encountering cross winds. If a cross wind force acts through the neutral steer point the car will maintain its heading and merely drift sideways; if the side force acts ahead of this point an unstable condition arises and ideally it should act slightly to the rear of the neutral steer point. Sharp corners on a body should be avoided as they produce sharp pressure change which can be aggravated by side gusts with consequent change in yawing moment. Figure 2.4 shows the force pattern; with centre of pressure at D at which air resistance T and normal wind force N act, a torque develops around the vertical axis Z through the centre of gravity S, namely, $M_s = N \cdot e$. Streamline cars for which $N \cdot e = (N_v + N_h)/2$ have a greater turning moment than bluff ones for which $N \cdot e = (N_v - N_h)/2$.

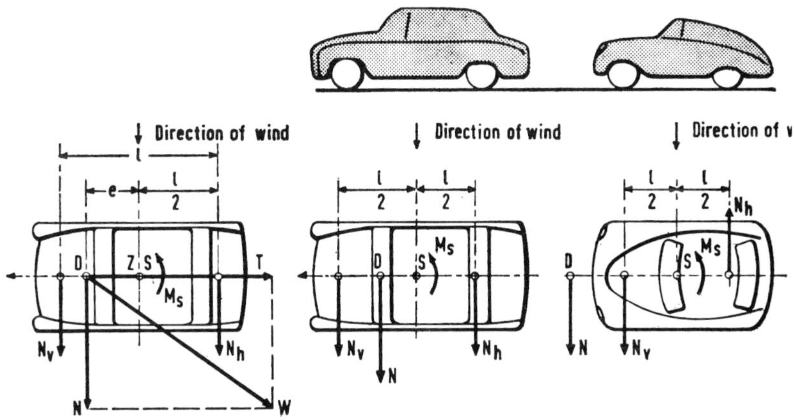

Fig. 2.4. Effect of cross winds on vehicle stability

EFFECTS OF MECHANICAL LAYOUT ON STRUCTURE DESIGN

The specialist car builder generally has the most scope for experimentation in mechanical layout because of the flexibility of his tooling, his fabrication skills, and above all the speciality of his market. One of the largest specialist builders, Lotus have succeeded by good engineering and marketing, in effectively bridging the gap between custom builder and quantity producer.

The Lotus Europa had a drag coefficient of only 0·30 and very low frontal area that made for one of the lowest air resistance figures in production car history. The packaging of a 'mid' engine installation and two adult occupants into such an interior envelope was an outstanding achievement. Figure 2.5 shows the car in diagrammatic side elevation and plan with key dimensions in inches. Vehicle (dry) weight is only 1350 lb (613 kg) using the 203 lb (92·2 kg) Renault 16 power unit. Fore–aft weight distribution is well balanced at 45–55 per cent.

Chassis structure is just detectable in the plan view being of central box backbone form, with fork extensions. It is a mild steel fabrication with substantial cruciform arms bifurcating at the rear to embrace engine and transmission, and at the front to carry a transverse box for suspension anchorages. Torsional stiffness of the basic spinal box is around 5000 lbf ft (6780 Nm)/deg but including the forked extensions this figure falls to around 1400 lbf (1893 Nm)/deg; a separate glass reinforced plastic (grp) body is fitted.

Although the mid engine position does provide exceptionally good weight distribution it obviously can cause packaging problems for the occupants and so in the family car, in particular, front or rear layouts are

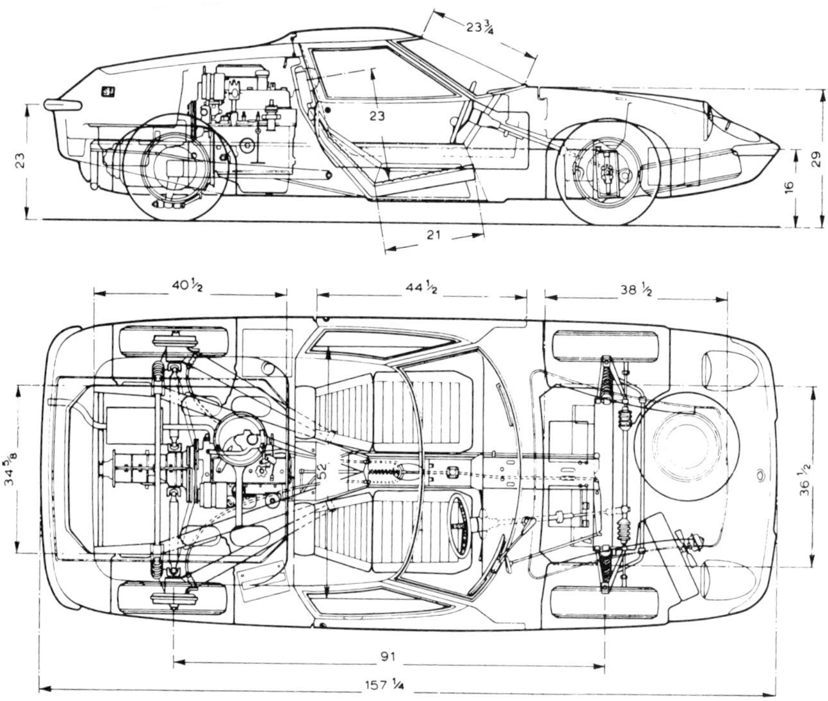

Fig. 2.5. Lotus Europa, elevation and plan

preferred. The arguments for and against grouped engine and drive layouts, front or rear, and the conventional front-engine/ rear-drive alternative, are more or less as follows.

Some say that because of the large clearance envelopes swept out by the steered front wheels, space between them is comparatively useless for other purposes and therefore should be used for the engine. Basic stability in handling too, follows from a concentration of weight at the front. The realization of good cooling and engine access, and generally a better gear shift mechanism came with rear drive. But front grouping is preferred for proximity of the driver to the primary engine controls and better structural integrity at the rear.

Those who favour rear grouping of engine and drive see no restrictions on turning circle by constriction of steered wheel swept clearances, better traction on slippery gradients, and a minimal change in centre of gravity position for the unladen and laden car. Noise, fumes, heat, and exhaust are all more easily removed together with the problem of steering the otherwise driven wheels. With the proximity of engine and drive, transmission power losses can be reduced and there is an overall occupant packaging advantage of leaving the space between the wheels clear for passengers.

Figure 2.6 compares the front and rear grouped arrangements applying

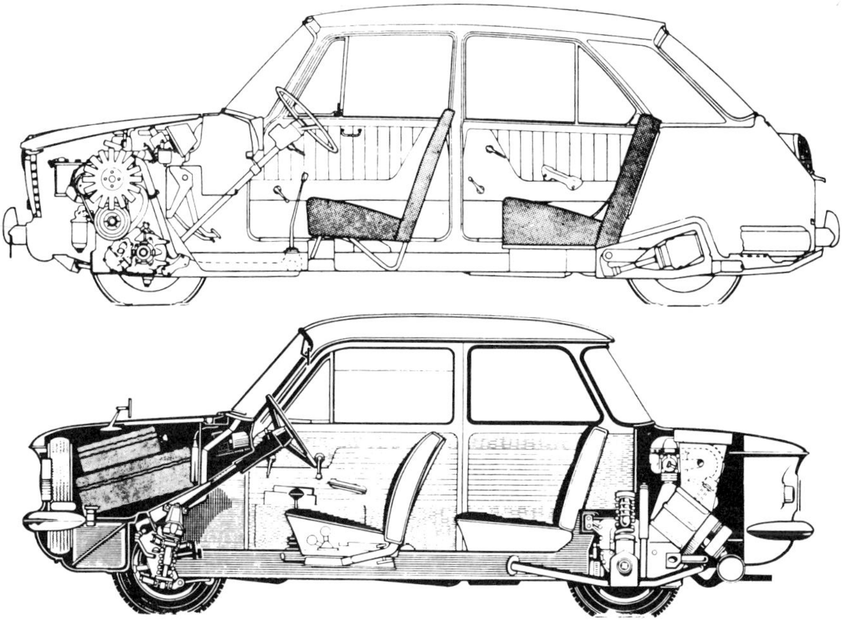

Fig. 2.6. Package comparison, BLMC 1100 and NSU Prinz

to the BLMC 1100 and NSU Prinz cars; the former has of course the added packaging advantage of a transverse engine so ample leg room and reasonable luggage accommodation are provided in an overall length of less than 12 ft 3 in (3·74 m). In the rear grouped engine/drive of the Alpine Renault A310 in Fig. 2.7 a 2 + 2 layout has been achieved and the structure, like the Lotus Europa, is a large tubular backbone. The car weighs 1818 lb (825 kg) and has 40–60 per cent front to rear weight distribution.

Sports car packaging

As profiles become lower in the sports racing and pure racing categories, and mid engined layouts are preferred, we see in the case of the Italian Gingiaro Manta and BRM of around 1970 (Fig. 2.8) that space for more than two occupants is almost impossible to achieve without an unreasonably high overall length. Whereas Marcos in Fig. 2.9 uses the conventional front engine rear drive arrangement the structural material is somewhat unusual in being a $\frac{1}{8}$ in (3 mm) thick marine plywood. Hull weight is a remarkable $1\frac{1}{2}$–2 cwt (75–100 kg) and shows what can be achieved with high specific rigidity materials. Main bending loads are carried by a central box beam which encloses the propeller shaft, and $13\frac{1}{2}$ in (343 mm) deep triangular-section box side members stabilized by

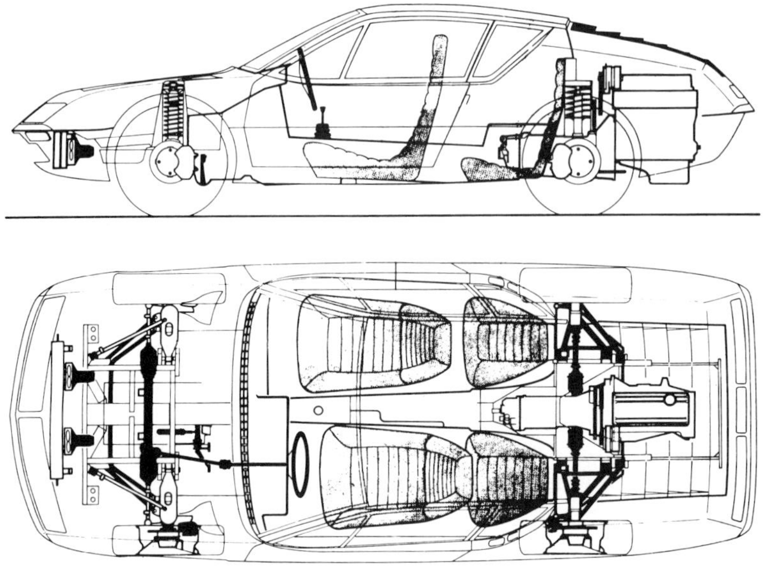

Fig. 2.7. Alpine Renault A310, elevation and plan

$\frac{3}{8}$ in (9·55 mm) thick bulkheads and intermediate diaphragms of $\frac{1}{8}$ in (3 mm) ply, reinforced by $\frac{1}{2} \times \frac{1}{2}$ in (12·8 × 12·8 mm) edge stiffeners. Side members are connected to the rear end box structure by a $\frac{3}{4}$ in (19·05 mm) thick multi-ply member. The timber structure terminates about 10 in (254 mm) behind the front wheels and a separate bolt-on sub frame of metal tubes is used to anchor the front suspension.

An example of a mid-engined car that does not rely on a central back-

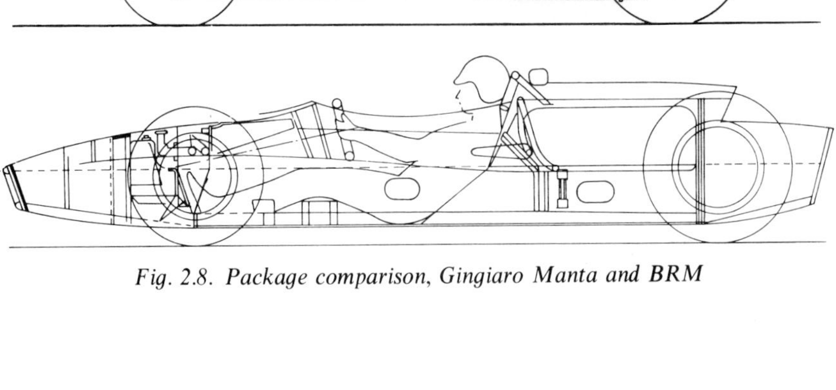

Fig. 2.8. Package comparison, Gingiaro Manta and BRM

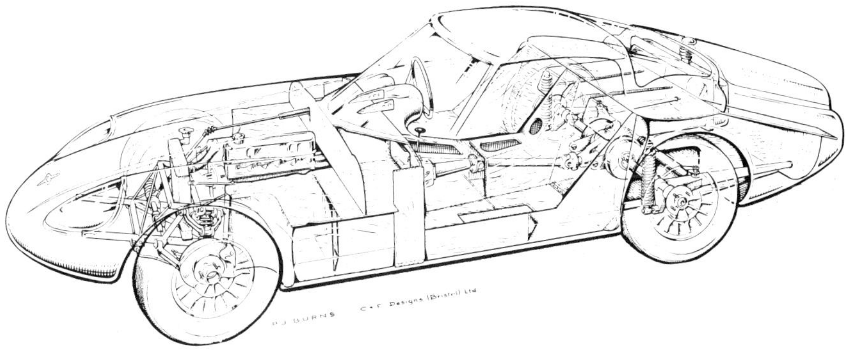

Fig. 2.9. Marcos, ghosted view of structure

bone structure is the Ford GT 40, which is just 40 in (1·017 m) high. The body–chassis is an integral structure, Fig. 2.10, built in thin sheet steel, 24–28 gauge. The underbody has deep box sills which house the fuel tanks; then there are two main bulkheads, a roof section and pick-up points for the suspensions. Torsional rigidity is 10 000 lbf ft/deg (13 560 Nm/deg).

Family car packaging

The Renault in Fig. 2.11 is seen to have a punt-like substructure. It and the superstructure weigh 397 lb (180 kg) which is 18·3 per cent of the total unladen weight. Noteworthy is the construction of the roof and body-side top member, the joint being capped by a chromed trim strip. The heater duct below the windscreen forms an additional stiffening member, and the double skinned rear quarter members embody the air outlets.

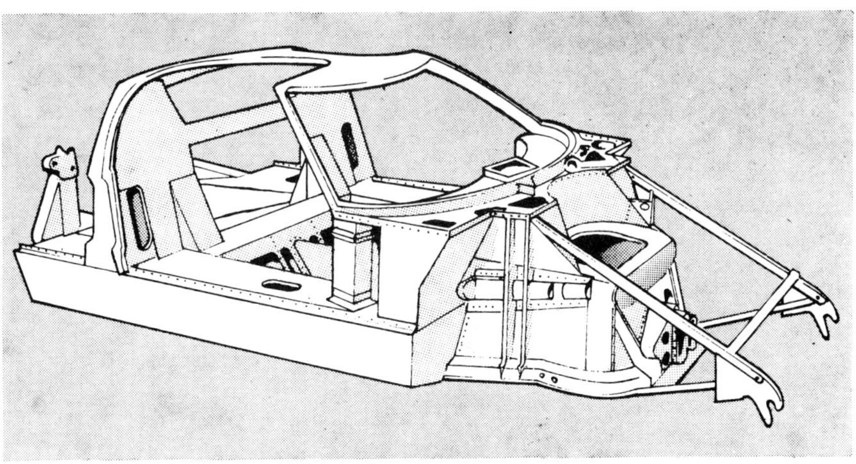

Fig. 2.10. Ford GT 40, perspective view of structure

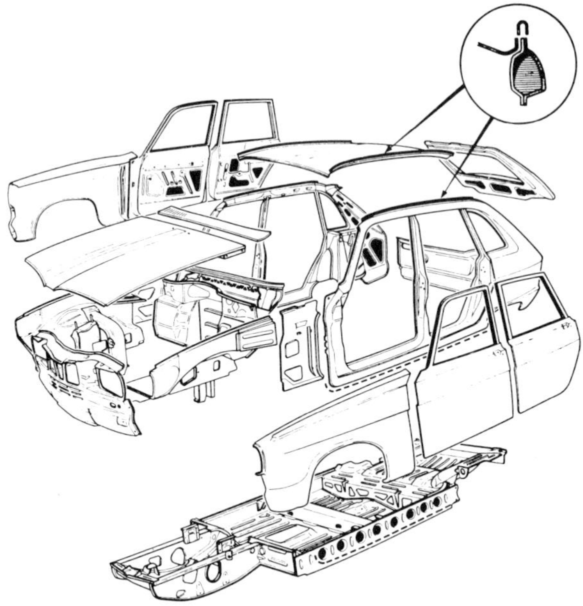

Fig. 2.11. Renault 16, exploded view of structure

The American Chevrolet Camaro in Fig. 2.12 reveals yet another structural solution to a packaging problem in the use of extensive subframes soft mounted to the main body shell. A number of contemporary American cars have gone yet further than this in the adoption of, so-called, perimeter frames, again soft mounted to the body shell. While the main aim of subframes and perimeter frames is to keep the body in a separate vibrational environment to the running units there are also other advantages, particularly in perimeter frames. Having a frame that runs around the periphery of the body shell underside means that all the running units are in a unitized assembly which can be fitted to different body types and even assembled in different plants from the bodies.

With conventional front engine rear drive configurations there is a critical packaging problem, particularly in low height sports saloons, in the area of the rear seat and drive axle differential unit. This problem is overcome on the Ford Capri, Fig. 2.13, by the use of short rear axle radius rods which rotate the axle at full bump to ensure that the nose of the pinion clears the rear seat profile.

In an experimental Ford project, the Techna car (see SAE paper 690 267) a different approach again is applied to drive line arrangement, an offset and canted configuration. Figure 2.14 compares the drive layout in a contemporary vehicle with that of Techna with the considerable gains in occupant space made possible without increasing the low silhouette of the

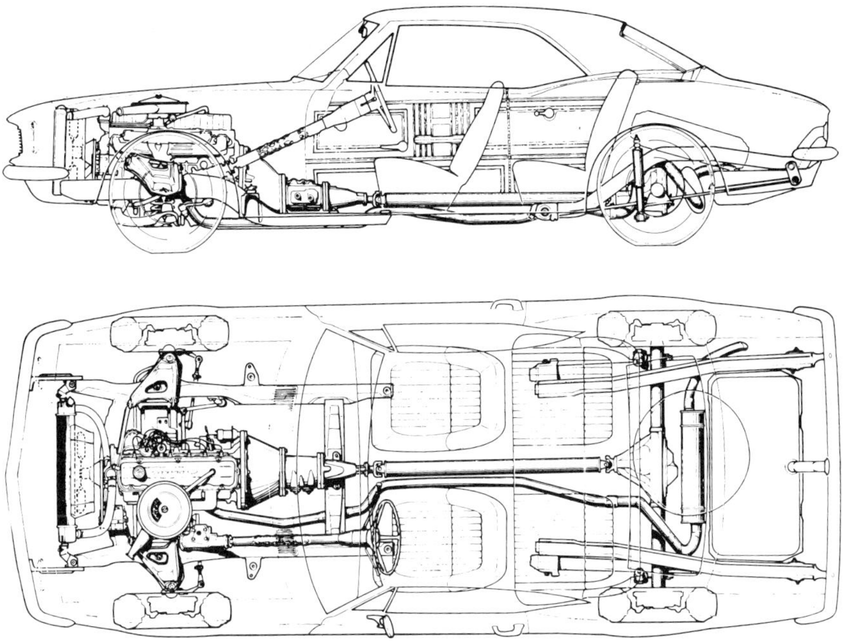

Fig. 2.12. Chevrolet Camaro, elevation and plan

vehicle. The solution complied with the usual constraints of economical manufacture and also involved forward location of the engine–gearbox.

Other ideas tried on Techna included bonded roof and floor panels, the latter being a honeycombed steel sandwich requiring much smaller sills than usual. This in conjunction with power operated parallel-link hinge doors, provides good access even in closely parked conditions. A 'structural glass' windscreen was also bonded in and hardly required corner pillars.

Special purpose vehicles

Moving into the area of cross-country vehicles as exemplified by the Range Rover we find that marked structural change takes place from the typical integral car structure. The body structure, Fig. 2.15, is attached to the chassis by ten rubber mounts plus four additional supports for the corrugated aluminium rear floor. Wings, also in aluminium, are of the bolt-on type and the aluminium roof panel is stiffened by longitudinal swages in addition to internal transverse hoop stiffeners.

The aluminium panels are connected to a steel framework, section through the sill of which is shown at (a), while at (b) can be seen A– and B–C-post sections, with at (c) the novel form of door latch. The two piece

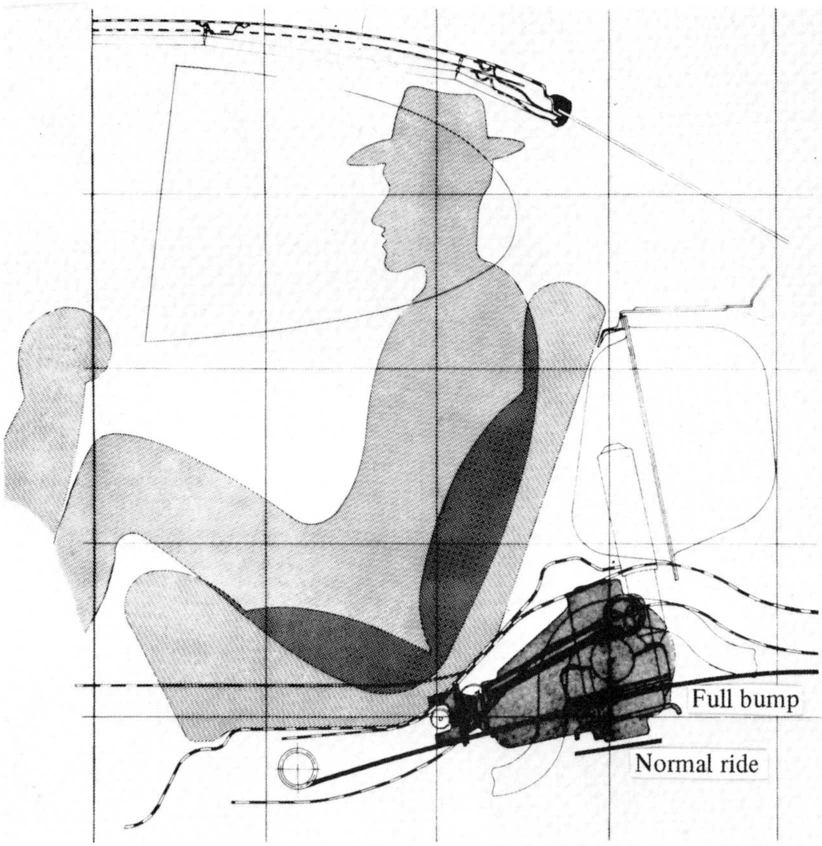

Fig. 2.13. Ford Capri, differential unit clearance

horizontally hinged tailgate is also in steel and is favoured for its ability to provide an extended deck for long loads (d). The deck capacity is 750 lb (390 kg) in addition to a full complement of five passengers. The gross vehicle weight is 5300 lb (2400 kg) and kerb weight 3800 lb (1725 kg).

The special position of goods vehicles with their separate chassis construction and multiplicity of body types will be considered in the final chapter. At this point it will be of interest to see how the material suppliers see future commercial vehicle trends.

In Fig. 2.16 is shown a fleet car/van/pick-up designed by ALCOA Aluminium to examine possible applications for its range of extruded sections and other fabrication processes. In the figure chassis D is of roll-formed construction, foam-filled. An extruded tailboard C folds flush with the deck for easy loading. At A a storage area exists for the sliding tilt and B shows the deck floor made from interlocking extruded planks. For the seats a

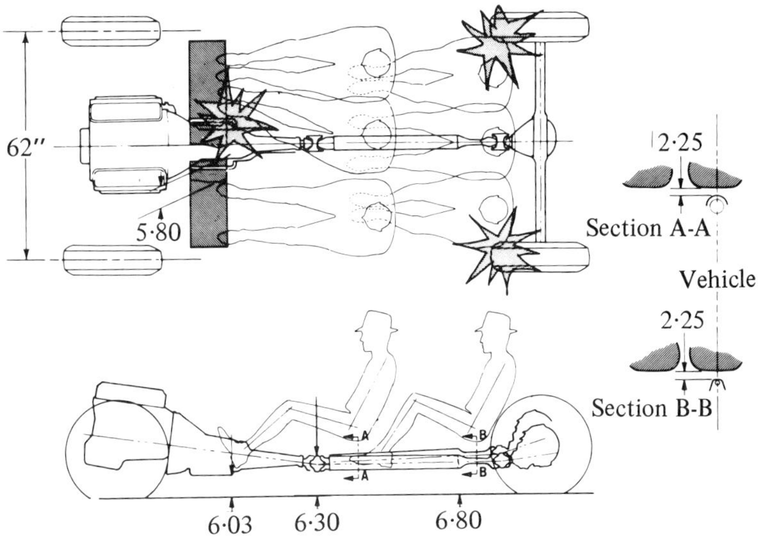

(a)

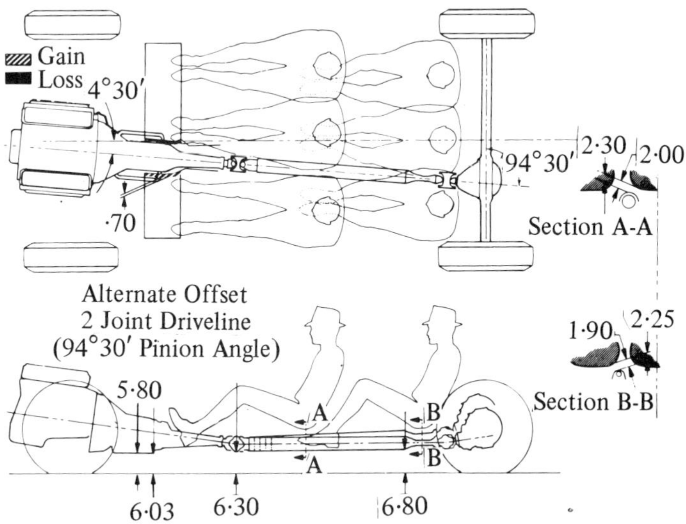

(b)

Fig. 2.14. Ford Techna, drive line clearance. (a) Toeboard space condition with conventional on-centre driveline and engine location (b) Techna alternate driveline

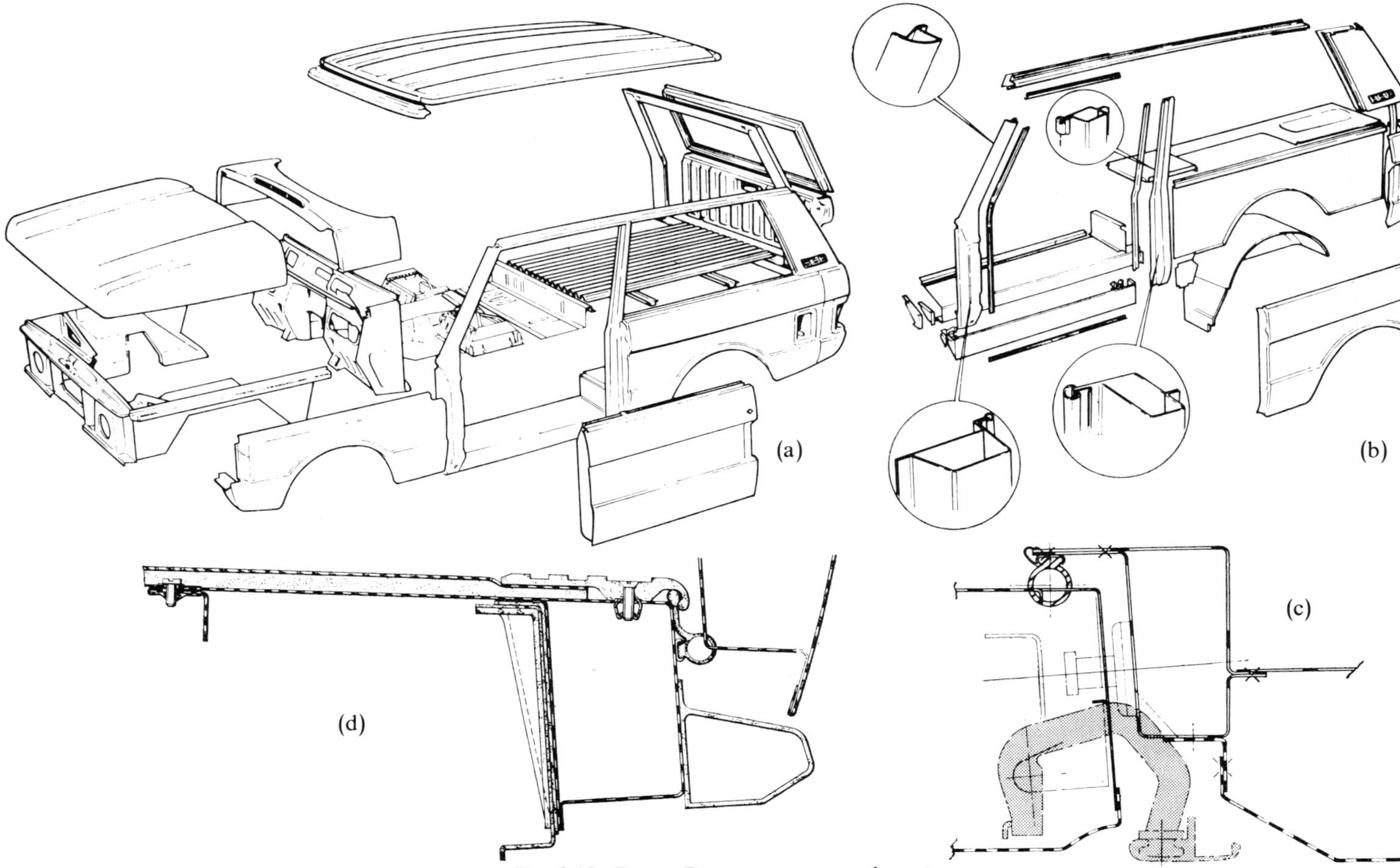

Fig. 2.15. Range Rover, structure and sections

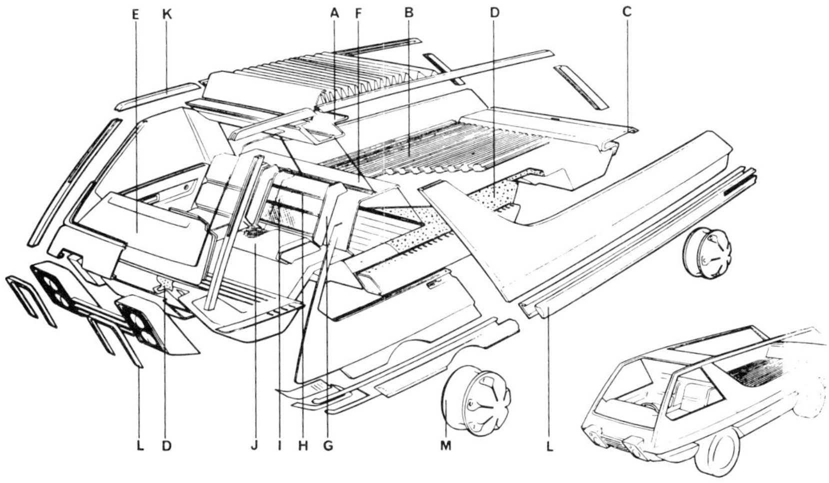

Fig. 2.16. ALCOA convertible van proposal

tubular structure H butts a cast frame G and supports aluminium screen mesh I around which foam rubber padding and fabric upholstery is bonded. F is a storage shelf, E a stamped instrument panel, and K an extruded screen frame. Wheels M are forged or cast and L signifies rubber inserts at critical contact points.

To promote the use of a variety of new rolled sections, US Steel have designed the multi-convertible 'Venture Wagon'. A horizontally opposed

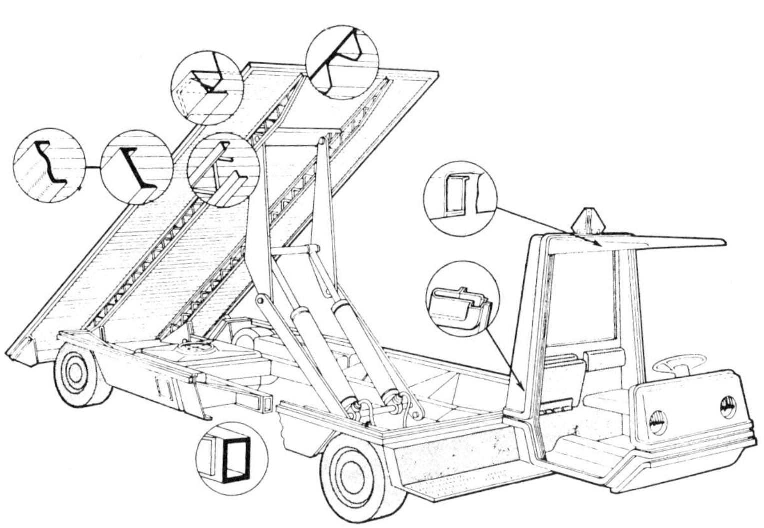

Fig. 2.17. US Steel Venture Wagon

engine is fitted in unit with the rear drive. Note, too the offset cab on an outrigger on the front left of the admirably low slung main and perimeter frame; this facilitates the carriage of long and awkward shapes, Fig. 2.17.

THE INTERIOR PACKAGE—PLANNING FOR THE OCCUPANTS

One of the most useful assets in the initial design stage of a motor car is a scale model average man which can be used to estimate the initial dimensions of the interior space whether one is designing to a full inclined racing car driving position or a taxi body for six passengers sitting upright. Figure 2.18 is a true scale model, each square representing a 10 inch grid at full size. By scaling up appropriately, this could be traced on to card, enabling the construction of a fully articulating cut-out man. This should prove useful as an adjustable template for drawing in drivers and passengers on to a 'package' layout.

The science of matching a machine (in this case a motor vehicle) to a man is called 'ergonomics'. A common word used by ergonomists is the 'percentile' which is used to describe the stature of a man or woman according to his or her place in the graph of percentage of population against overall stature dimensions. It is usual to design within percentile extremes on the basis that there are very few women, say, smaller than the 5th percentile and very few men larger than the 95th, these corresponding,

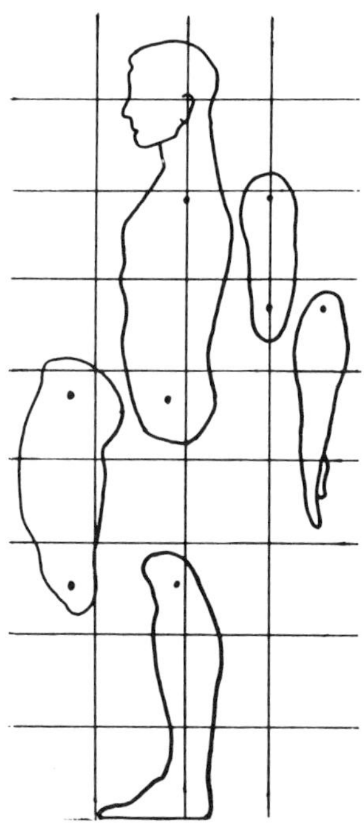

Fig. 2.18. Average man, fitting template

respectively, to 58 and 72 inch statures. Hywell Murrell writing in *Automotive Design Engineering* for October 1963 shows in Table 2.1 and Fig. 2.19 the factoring of data to accommodate the percentile variations. He goes on to show how the vehicle controls can be fitted to the driver.

Foot controls

Foot controls are divided between those like brakes, needing considerable force, for which free leg movement is required, and the 'precise' control of throttle, say, where use of ankle articulation is better suited. For braking and declutching a situation of progressive force with virtually zero movement is ideal.

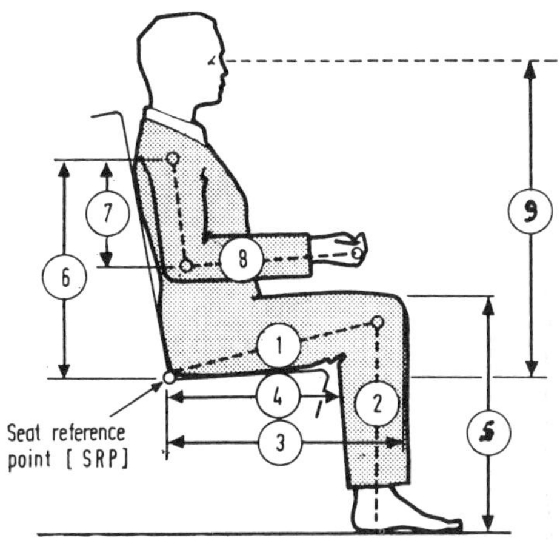

Fig. 2.19. Key dimensions, seated man

TABLE 2.1

Dimension	Percentile male			Percentile female		
	5th	50th	95th	5th	50th	95th
1 SRP to knee hinge	20	22	$23\frac{1}{2}$	$18\frac{1}{2}$	$20\frac{1}{2}$	$22\frac{1}{2}$
2 Knee hinge to sole of foot	$18\frac{1}{2}$	20	22	$15\frac{1}{2}$	18	20
3 Buttock to patella	$21\frac{1}{2}$	$23\frac{1}{2}$	25	$20\frac{1}{2}$	22	24
4 Buttock to back of knee	17	19	$20\frac{1}{2}$	16	18	$19\frac{1}{2}$
5 Knee height (no shoes)	20	$21\frac{1}{2}$	$23\frac{1}{2}$	17	$19\frac{1}{2}$	$21\frac{1}{2}$

Dimension	*5th %ile female*	*95th %ile male*
6 Shoulder point above seat (sitting shoulder height less 2 in)	17 in	23 in
7 Shoulder point to elbow hinge (shoulder to elbow less $1\frac{1}{2}$ in)	$10\frac{1}{2}$ in	$14\frac{1}{2}$ in
8 Elbow hinge to closed hand (elbow to fingertip less 1 in for hinge and 4 in for closed hand)	11 in	15 in
9 Sitting eye level above seat	$25\frac{1}{2}$ in	$33\frac{1}{2}$ in

Critical factors for pedal force and movement are: knee angle in leg operation optimum 160–165 degrees, cf. most comfortable relaxed range 120–130 degrees, and ankle angle in leg and ankle operation, within 90–125 degrees and 75 degrees absolute minimum. Both angles are obtained from a spatial relationship of seat and control; see Figs. 2.20–2.23. Optimum lateral offset of pedals from body midline is 2–5 in (50–130 mm); at 6·5 in (165 mm) offset, leg force potential is reduced by 75 per cent.

Alternative approaches in calculation are either to determine seating/pedal position, given the force required hence knee angle from Fig. 2.20; or given seat/pedal position to determine maximum allowable force. Optimum seat parameters are shown in Fig. 2.21 and Table 2.2.

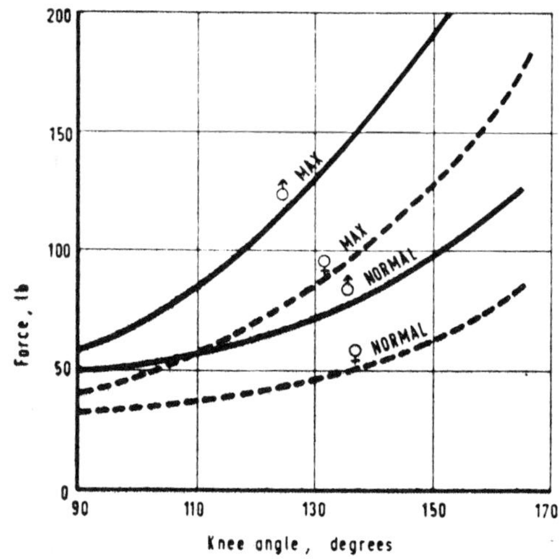

Fig. 2.20. Maximum (instantaneous) and normal (repetitive) forces that can be applied by the leg

A worked example to find leg operation parameters for 95th percentile man to 5th percentile woman is as follows, using the data from Fig. 2.22. Procedure for calculation is to design for adequate vision of the shortest woman and then use available pedal pressure for this position from Table 2.3. Pedal movement is then arranged to be in line with the lower leg. The greatest force can be applied by the instep; it is reduced by 50 per cent if toes are used. Hence it is best to site the pedal 6 in (150 mm) above floor level with further 2 in (50 mm) clearance for toes. And pedal travel should be restricted to 2 in (50 mm) maximum before pressure is required.

Ankle controls

Ankle controls like throttle are best with pedals hinged at their base.

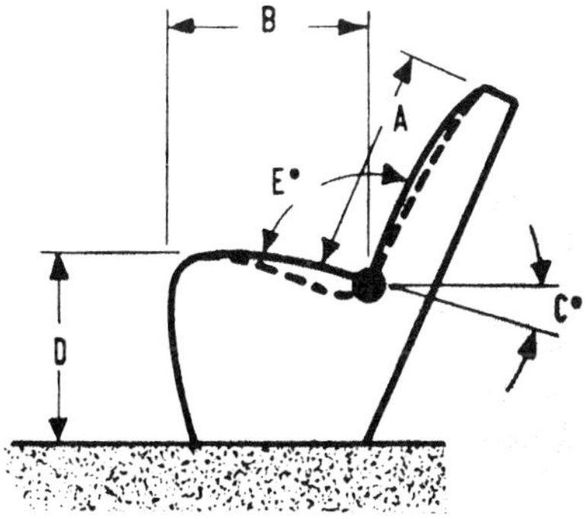

Fig. 2.21. Seat parameters

TABLE 2.2

SEAT PARAMETERS
Optimum E = 105° (90% acceptable, 102°–108°)
C = 6° (90% acceptable, 5°–8°)
B ≤ 16 in for small women
= 17 in as compromise (provided knee pivot of small woman > $2\frac{1}{2}$ in above seat front edge, to allow full pedal depression)
A = 18–24 in
D < 15–16 in

* Failure to meet this dimension causes accident risk due to reduced pedal force arising from popliteal pressure. Seat front edge must be rounded and softly padded.

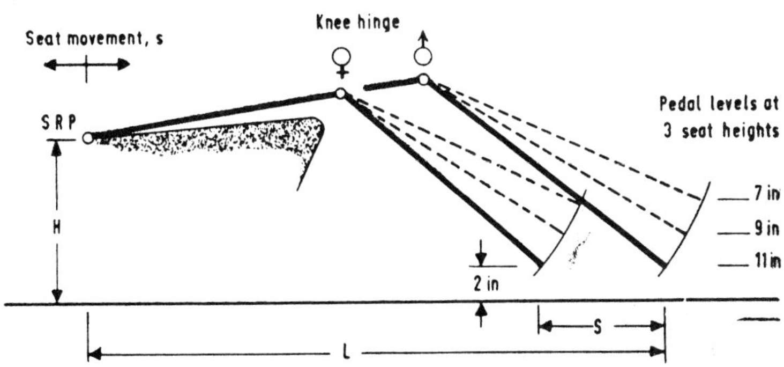

Fig. 2.22. Leg control reach envelope

TABLE 2.3

Normal and maximum (emergency) pedal efforts given by shortest women and tallest men in 90% range, for three seat heights; M = Males, F = Female.

Seat height (in)		*Knee angle*		*Norm. force* (lb)		*Max. force* (lb)	
SRP	Front edge	M	F	M	F	M	F
11	13	133°	126°	75	45	135	80
9	11	140°	142°	80	50	155	95
7	9	145°	135°	90	55	175	120

Optimum and maximum angles for barefoot are given by Fig. 2.23 and Table 2.4. When pedal is hinged at heel floor contact point, angle E should be increased, to allow for heels, by 5 degrees for men and 10 degrees for women compared with no shoes condition. For example, in Fig. 2.23, $E = 45$ degrees (55 degrees with shoes) for women and $E = 55$ degrees (58 degrees with shoes) for men (also $E = 71$ degrees and 73 degrees for men and women with 7 in (180 mm) seat height) so an acceptable pedal can be arranged.

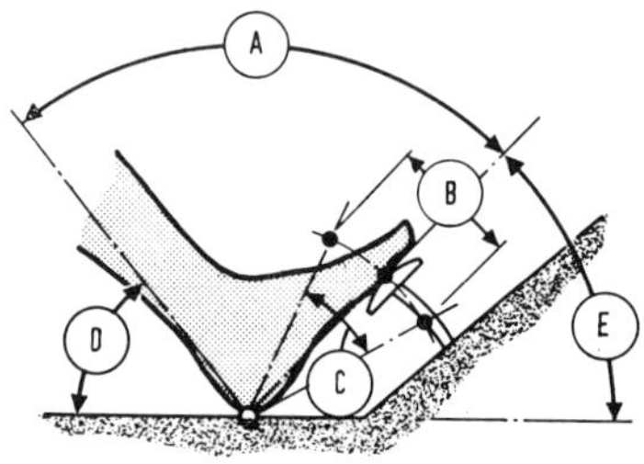

Fig. 2.23. Foot control parameters

TABLE 2.4

A optimum = 90°—'off' position
A maximum = 125°—'full on' position
A absolute minimum = 75° to avoid pain in toe extensor muscles
C max = 15° (corresponding to B max = 3 in)

Designing controls for ankle operation starts with the determination, from Table 2.1, of seat heights for population extremes and finding corresponding D in Fig. 2.23. A pedal angle is then selected to lie within allowable range. Throttle and brake pedal surfaces are then aligned for quick foot interchange, so the throttle pedal is mounted closer to the Seat Reference Point (SRP) than the depressed brake pedal (using the smaller knee angles than for the brake pedal when calculating the position). Sufficient throttle spring loading is required to support the foot (4–8 lbf). As an example, when $E = 55$ degrees to suit the smallest woman on an 11 in (280 mm) seat, A is 115 degrees for the largest man on a 7 in (180 mm) seat. If depression is 15 degrees, this gives an unacceptable maximum for A of 130 degrees. Remedy is to increase E to 60 degrees and reduce the depression.

Hand operated controls

Murrell next goes on to relate the areas of reach for these to the SRP as in Fig. 2.24; observation of these limiting dimensions is particularly for seat-belted occupants. Note that largest members of the population could be worse off than low percentiles in the reach situation; this is because they

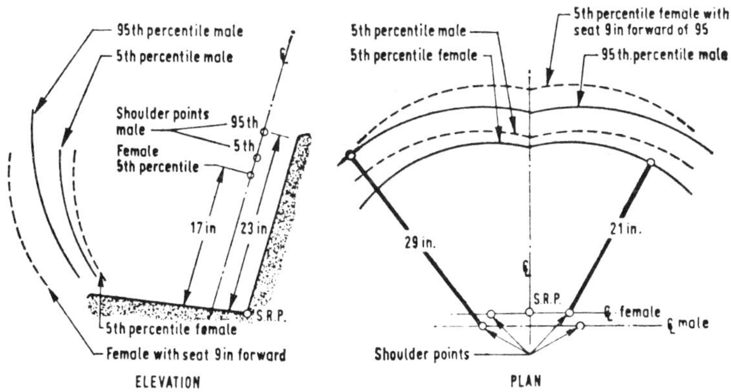

Fig. 2.24. Hand control reach envelopes

would normally be required to sit in a rearward position for satisfactory foot control. Steering column controls, particularly for quickly required items such as lamps and screenwipers, are valuable in allowing a remotely-placed fascia for safety in accidental impact.

For the steering wheel, applied hand force depends on column rake, perpendicular distance from SRP to wheel centre, and seat back inclination. Maximum force is available pulling longitudinally with forearm and shoulder acting together while faster, but lower efficiency motion is obtained with movement across the body. A wheel tilted 30 degrees to the vertical could have a rim force 1·7 times that with a vertical wheel but the rate of turning is halved and the energy expended reduced considerably.

A sports/racing car requiring accurate control is suited by a vertical wheel; an extended arm position gives maximum speed for what is a short time activity of high stress. Heavy commercial vehicles with high steering load, when power assistance is not available, require an almost horizontal plane, though this is changing with the standardization of power steering. For private cars 40–60 degrees wheel inclination to the horizontal is satisfactory.

In Fig. 2.25 angle A should be from 0 to 40 degrees, B between 80 and 125 degrees, C between 14 and 19 degrees, D greater than 6 in (150 mm), and 8 in (200 mm) for $E = 12$ in (300 mm) and 15 in (380 mm), respectively, and F ideally equal to 14 in (350 mm). These dimensions are based on hands gripping the wheel at extreme width on a horizontal axis.

Design procedure is to locate wheel centre and inclination for population extremes and determine reach envelopes described by closed fists using dimensions in Table 2.1 and Figs 2.19 and 2.24. Wheel centre must then fall in common area as shown by the example in Fig. 2.26. If column length is non-adjustable no position meets population extremes; for a 45 degree wheel the small area X applies and for a 30 degree wheel the larger area Y is common.

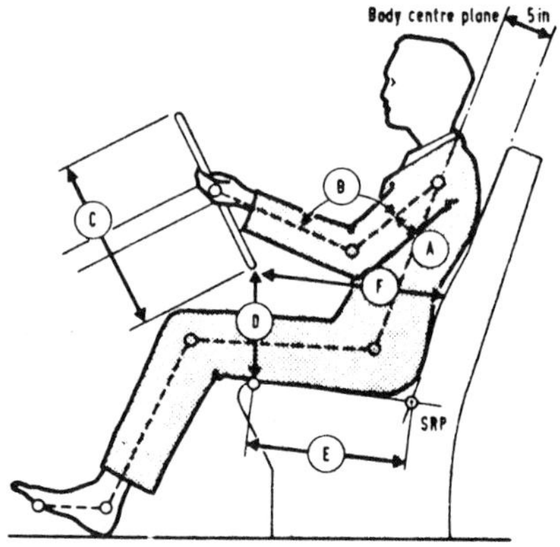

Fig. 2.25. Key dimensions, hand controls

If column length is adjustable by 3·5 in (90 mm) for this example then a satisfactory condition exists for both driver extremes, anywhere with a 30 degree wheel and for a small range within the 45 degree inclination. Figure 2.26 shows a range of positions based on the wheel centre being at the lowest point in the woman's area. Figure 2.27 shows the resulting improved woman's vision when an adjustable column is used.

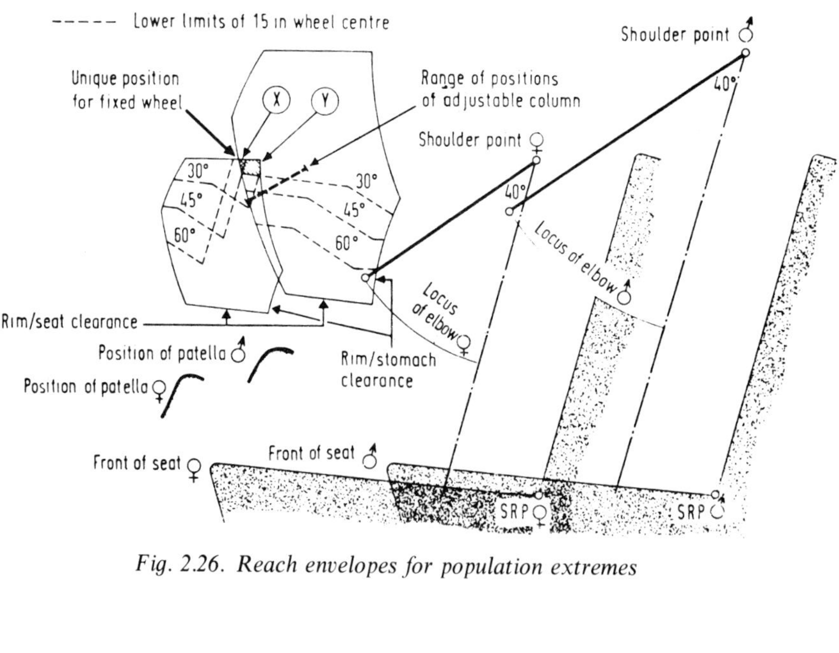

Fig. 2.26. Reach envelopes for population extremes

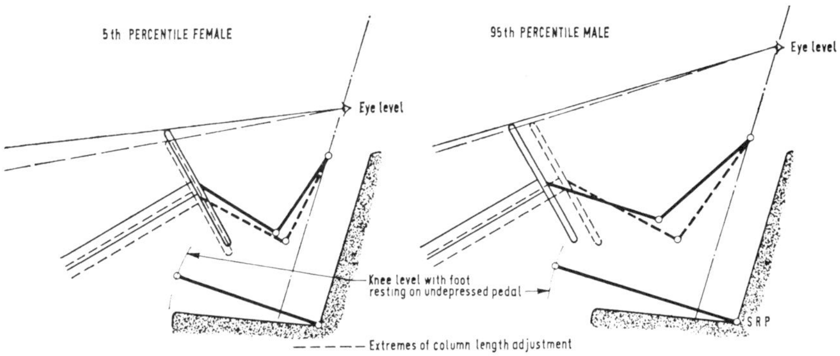

Fig. 2.27. Improved vision with adjustable steering column

CASE STUDY IN VEHICLE LAYOUT AND DESIGN

Body structures are not merely affected by the needs of mechanical unit configuration and occupant packaging; the use to which the vehicle is put makes constraints on the body structure as do the needs of the manufacturing technique employed.

Perhaps the best of all vehicle categories for examining external constraints on design is the bus. The following is an examination of the factors leading to the structural analysis of the Leyland National bus, from a

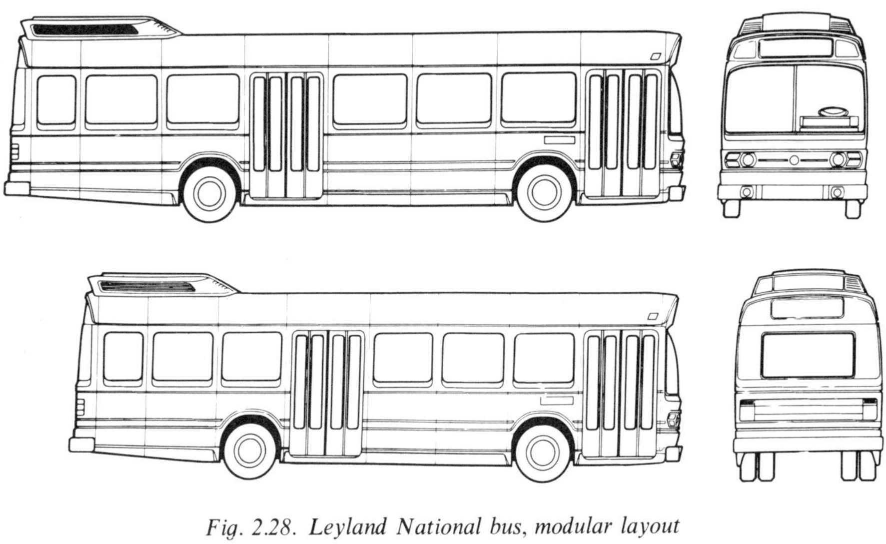

Fig. 2.28. Leyland National bus, modular layout

paper by McGowan and Hemmings read to the Institute of Body Engineers in 1973.

The bus was to be a single-deck rear-engined city vehicle to be marketed in two basic lengths, Fig. 2.28, designed for left- and right-hand drive, with single or double doorway configuration. Generally the philosophy influencing the structural design could be reduced to the effects of international markets, social advancement, operating requirement, production, and vehicle system requirements.

International Markets

Aiming the vehicle at a broad export market involved complying with a wide diversity of Construction Regulations from country to country. A major facet of such legislation is safety design; the Leyland National had to be resistant to side-swipe, rollover, and extensive collapse on frontal impact.

The body and underframe was to be constructed from varying sizes of 15 ton/in^2 (10·94 kN/m^2) yield steel with the use of pressed steel cross members in the suspension areas and pressed steel wheel arch risers. Sideframing was to comprise formed longitudinal members with pressed steel door frames and areas adjacent to the wheel arches reinforced by steel truss panels.

The heating and ventilating system had to cater for extremes of temperature and humidity and yet have its air distribution ducting built into the structure of the bus without loss of interior space. At the same time the bus had to be comparable in terms of weight and price with conventional vehicles. Also the vehicle had to be built in short time so as to overcome the arresting of capital otherwise tied up in the lengthy build periods of conventional buses.

Modular construction permits the bus to be built in three different lengths to suit varying national requirements. Common structural areas are (a) front end including driver's compartment and front axle, (b) the centre door, rear axle and power pack, (c) the extreme rear end with cooling/heating equipment. These are interspersed with two sizes of similar body modules, or bays, which can be arranged as required between common designs of pillar and roof stick 'ring' framing, (Fig. 2.28).

Social advancement

The highest levels of performace, safety and comfort are required for passengers and driver—comfortable seating, smooth jerk-free ride, good vision, Fig. 2.29, low noise, good lighting, wide safe doorways, flat safe floors—all

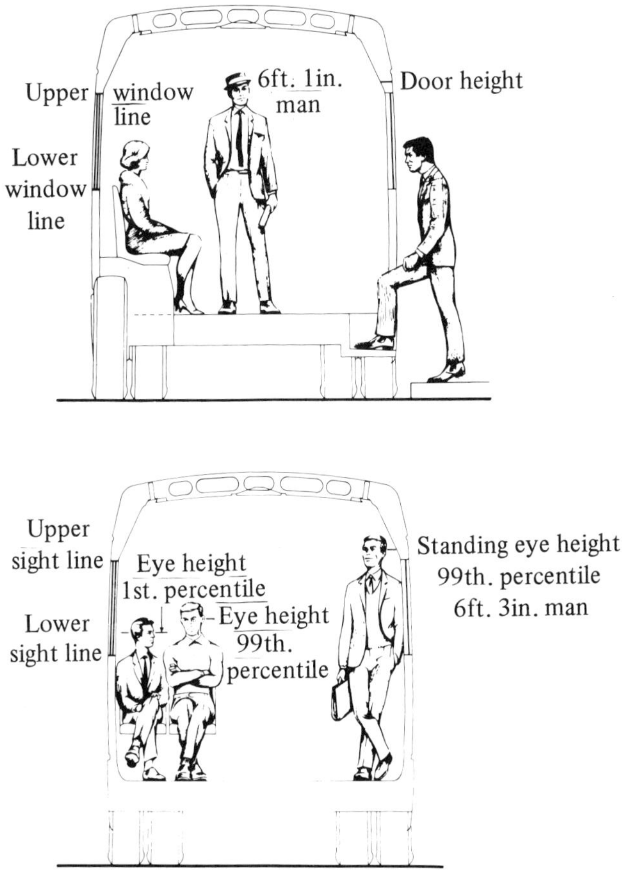

Fig. 2.29. Leyland National bus, passenger vision

these criteria important to passengers were to be met in the single design of bus.

The driver was to receive considerable attention with an environment tailored to accommodate the largest range of size and shape, Fig. 2.30. Driving effort was to be minimal, with controls carefully positioned. All this was to be contained in a structure giving him a high degree of safety in impact yet without restriction of vision.

Operating requirement

The bus was designed from the outset for one-man operation. Requirements for service staff also needed attention, with commonized parts for simple spares stocking, commonized layout for easy service training and

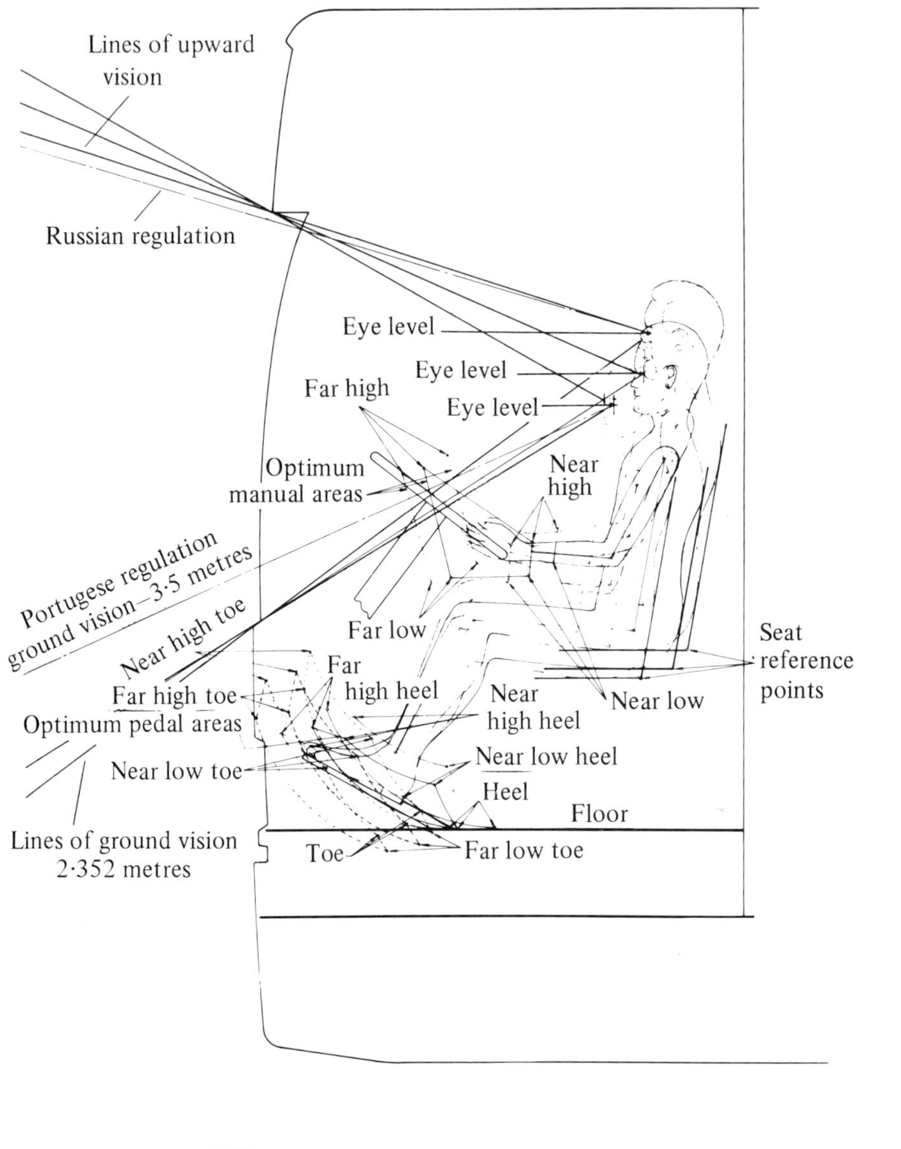

Fig. 2.30. Leyland National Bus, driver vision and reach envelopes

quick replacement of panels or areas. Unpainted trim on the inside and smooth exterior contours were to allow the use of mechanical cleansing equipment.

Production

Early on it was recognized that the cost of the vehicle could only be controlled by introducing semi-automated tooling and minimizing direct labour, the degree of tooling being directly related to the production volume. To maximize tooling utilization and reduce piece price a maximum quantity of standardized components had to be sought.

Body structural joints and elements had to be designed as much for production assembly technique as performance requirement. Use of press tooling and spot welding for large volumes of standard items would favourably affect cost as well as improve serviceability in the good dimensional control that could be obtained.

Main assemblies would be Avdel-riveted together, these comprising floor, roof, RH/LH sideframes, front and rear assemblies. The build up of these units and eventually the complete vehicle would be to a track discipline.

Vehicle system requirements

Suspension, heating/ventilating, door-gear and many other items impose limits on the structure. The layout of the vehicle shown in Fig. 2.31 is an indication of the need to design the structure for the complex load inputs. Even the transverse roof support member required apertures for the longitudinal ducting of heating air, for example.

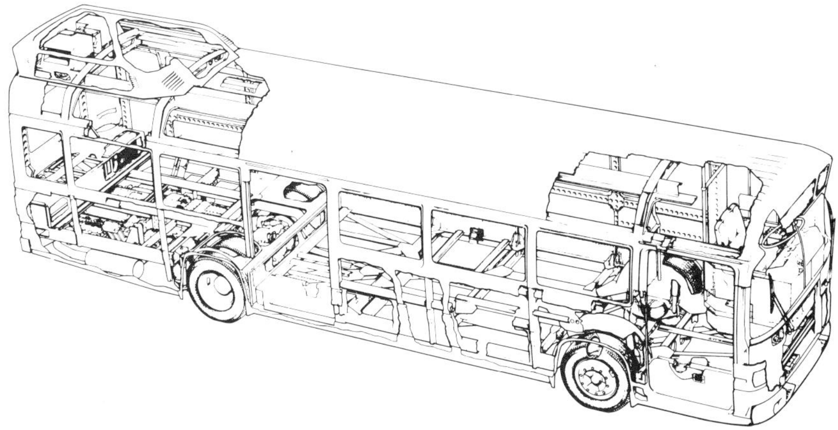

Fig. 2.31. Leyland National bus, structure and running units

Approach to the design

The basic information requiring definition, related to weight and cost targets, was as follows.

1. Vehicle package and dimensional layout.
2. Structural requirements: life and performance targets.
3. Production requirements: fastening and assembly techniques.
4. Material specifications.

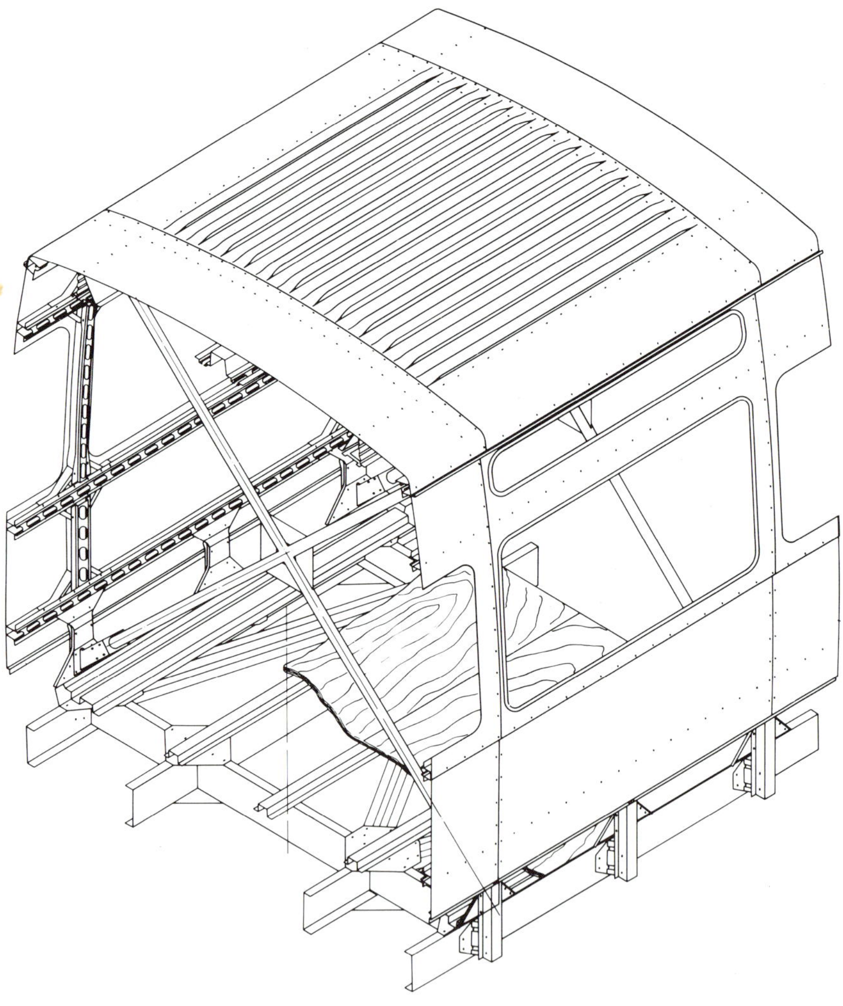

Fig. 2.32. Leyland National bus, structural module

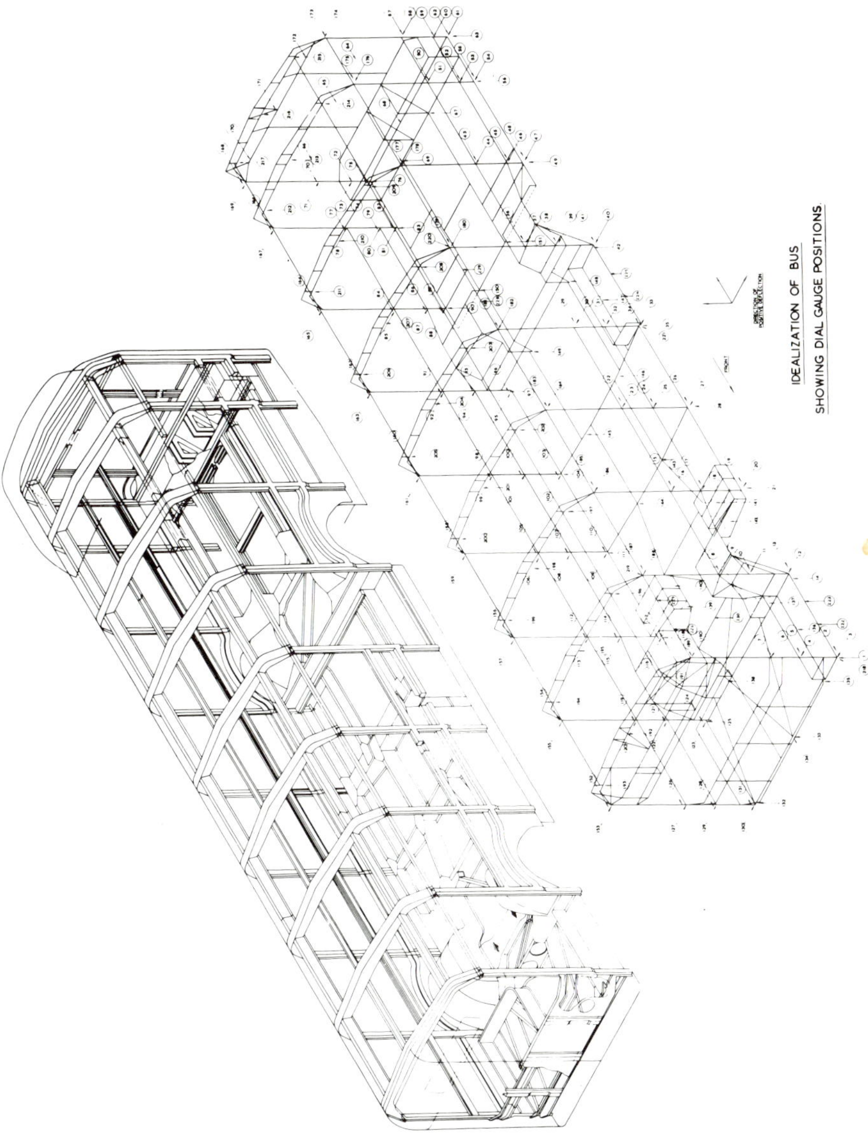

Fig. 2.33. Leyland National bus, structural framework and idealization

A life target of 15 years was fixed and it was decided that normal working stress in the components should be approximately one third of the yield stress, or 0·1 per cent proof stress, of the component material. The vehicle structure was to have at least twice the torsional and beam stiffness of current UK traditionally built single-deck PSV's. Also, it was the aim to balance out stress levels around the structure and in particular for the roof structure to be fully contributory over the doorways. Furthermore the structural performance was to be equal to the standards laid down by the American Federal School Bus Roof Load Test, the West German Passenger Load Design Criteria, and the Swedish Protection Around the Driver Compartment requirements. Finally, a requirement was set to offer occupant protection in impact by arranging the structure to absorb energy during a controlled deformation.

The overall loading criteria included 3*g* vertical intensity, in some areas 4*g*; 2*g* vertical plus horizontal braking, forward and reverse; 2*g* vertical loading plus left- and right-hand 1*g* radial loading; and a torsional load input of 50 000 lbf ft (67 700 Nm) at the axles.

Fastening technique was to be spot welding for all suitable and economic sizes of sub assembly and Avdel riveting together of sub- and main-assemblies. Materials chosen were low alloy steels for major structural elements and aluminium for lightly loaded components. There would be liberal use of hot dipped and electroplated galvanized steels in the underframe, as well as epoxy-powder sprayed members.

Analytical approach

Individual elements and joints such as pillars, transverse roof supports, cross members, wheel arch risers and underframe joints were first analysed, tested, and structural/production solutions obtained. Next stage was the evaluation of large structural systems such as wheel arch groups, door frames, step and engine beam assemblies.

To test the analytical techniques for the complete structure a module of the bus Fig. 2.32 was first examined and idealized for analysis. It was possible to predict the torsional stiffness of this structure to within 25 per cent. At the same time techniques for predicting the stability of the stressed panels, using aeronautical design techniques, were applied prior to the analysis of the complete structure. The latter was analysed by a three-dimensional programme involving finite-element techniques; the idealization is shown in Fig. 2.33.

CHAPTER 3

BEAM ELEMENTS, THE EQUATION OF FLEXURE, STRESSES, AND DEFLECTIONS

SHEET METAL REPRESENTATION

WITH the obvious exceptions of lightweight truck, bus, and performance car bodies, vehicle structures (particularly the quantity-built car) are made from extra deep drawing mild steel to EN 2A/1 with an ultimate tensile strength (UTS) of 45 000 lbf/in^2 (310×10^6 N/m^2) and 0·5 per cent proof stress between 27 000–38 000 lbf/in^2 ($186–263 \times 10^6$ N/m^2). Work hardening in forming structural elements can raise UTS to 53 600 lbf/in^2 (371×10^6 N/m^2) and 0·5 per cent proof stress to 45 000 lbf/in^2 (310×10^6 N/m^2). Modulus of elasticity is 30×10^6 lbf/in^2 (210×10^9 N/m^2) for most steel.

Minimum sheet-metal gauges are set by considerations of in-factory handling and susceptibility to damage. Above these values, gauges of loaded members will be fixed to ensure resistance to compressive buckling or shear instability. In so-called 'thin-walled structures', a simplified analysis can be based on ignoring all loads on a strained element with respect to which the thin wall is flexible; on that basis panels carry *only* shear and tension (this leads to complementary shear flows (see page 57) in the wall being equal).

Flat sheet panels can be made resistive to end-loading by the introduction of swages. The axes of swages should be straight and swages should run preferably along the shortest distance between supported edges of a panel. The stiffness of the swage depends mainly on its depth and swages should not intersect since loss of stiffness occurs at the point of intersection. Unsupported panel edges can similarly be flanged to obtain end-load capacity. Swages and flanges can be idealized as end-load carrying channels.

J. Pawlowski in his book, *Vehicle Body Engineering*, Business Books, 1969, bases a whole system of vehicle stress analysis methods on the assumption of end-load carrying members and adjacent shear panels idealizing real structural elements. He defines a structural surface on a vehicle, as one in which all forces act in the plane of the surface, which may be a panel or framework rigid in its own plane. A shear panel structural surface

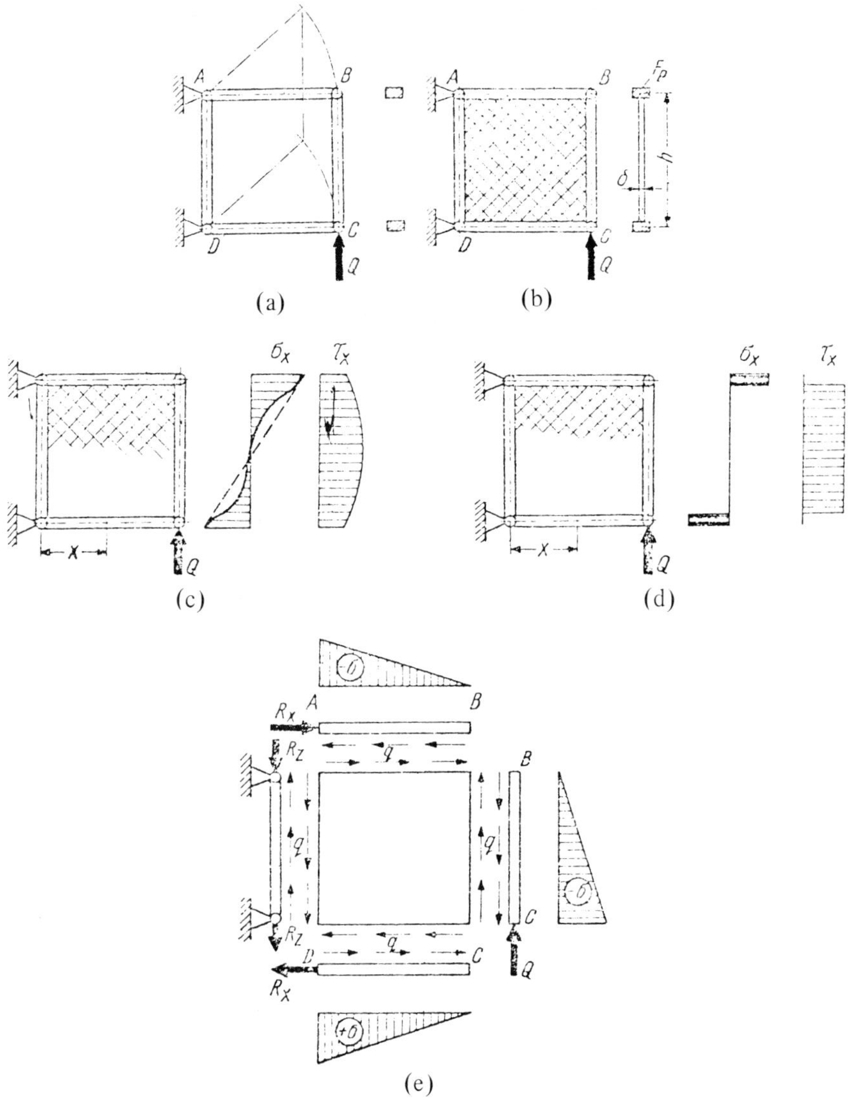

Fig. 3.1. Idealization of a shear panel

idealization is shown in Fig. 3.1. At (a) a flexible pin-jointed framework becomes rigid, by the addition of a shear panel\(b), with respect to force Q. The actual stress distribution is shown at (c) and the one assumed for the idealization at (d), where the edge members carry all direct stress and shear stress on the panel is constant. The equilibrium conditions and stress distribution assumed on the bars are shown at (e).

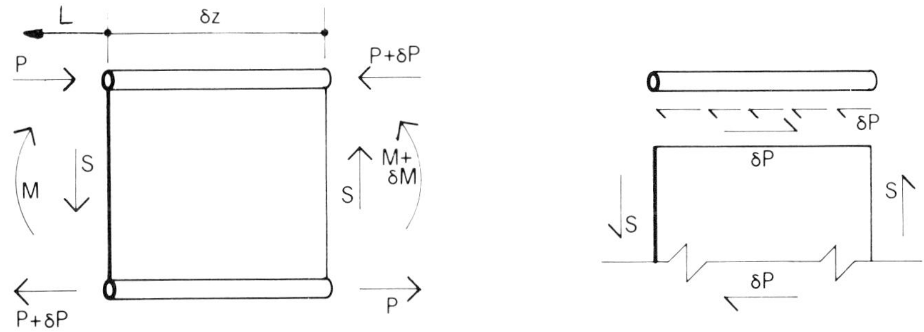

Fig. 3.2. Element of idealized thin-walled beam

The co-operation of the real structural skin in the shear panel method is accounted for by the addition of 20–60 × sheet thickness as an additional 'width' of material acting with the bar member at the edge of the shear panel. The magnitude of pure shear stress in the panel

$$\tau = Q/ht$$

for thickness t and shear force Q. Normally shear is expressed in 'flow' terms (shear flow = shear force per unit length) so that shear flow

$$q = Q/h = \tau t$$

Under the action of bending moment, say, M in Fig. 3.2, longitudinal stress in the bar $\sigma = P/A_f$ where A_f is bar, or boom cross-section area. Therefore $\sigma = M/dA_f$ where d is distance apart of booms. Yet simple theory gives $\sigma = \{M(d/2)\}/I$ thus the effective second moment of area

$$I = A_f \cdot \frac{d^2}{2} = 2A_f \frac{(d)^2}{(2)} = \sum A_f \cdot \bar{y}^2$$

in conventional notation where $\bar{y}$ is the distance of the boom area from the neutral axis. In Fig. 3.2, at distance L along the idealized beam from the point of application of shear force S, end load in booms

$$P = \frac{M}{d} = \frac{SL}{d} = qL$$

for shear flow q. Over an elemental length, a 'distributed' force is applied by the flange to the web where

$$\delta P = \delta \frac{M}{d} = \frac{S\delta}{d} z$$

so

$$\frac{\mathrm{d}P}{\mathrm{d}z} = \frac{S}{d} = q$$

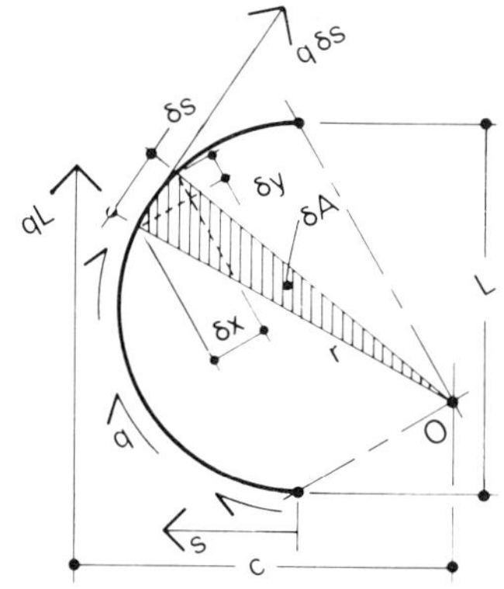

Fig. 3.3. Idealized beam section with curved web

Thus the shear flow in the 'web' is equal to the rate of change of end load in the boom. The assumption of constant shear flow in the web is made whether the web is straight or curved.

Curved panels

Peery, in his book *Aircraft Structures*, McGraw-Hill, '950, considers the generalized curved web shown in Fig. 3.3, to determine the resultant force from the shear flow in the web. The force on elemental arc δs is $q\ \delta s$ with components $q\ \delta x$ and $q\ \delta y$, horizontally and vertically. Integrating these components over the web gives total force components

$$F_x = \int_0^x q\ .\ \mathrm{d}x = qx \qquad \text{and} \qquad F_y = \int_0^y q\ .\ \mathrm{d}y = qy$$

with resultant qL. To find the moment arm of this resultant about a point 0, the moment arm of the element force $q\ \delta s$ is r and the area subtended by δs is $r\ \delta s/2$ so the total moment is

$$\int qr\ \mathrm{d}s = \int 2q\ .\ \mathrm{d}A = 2[A]q$$

where $[A]$ is the area enclosed by the lines joining the ends of the web to 0. Hence

$$c = 2[A]q/qL = 2[A]/L$$

In closed section idealized beams different shear flows are developed in shear panels enclosed by pairs of booms. These flows are in equilibrium with the externally applied load.

A simplified idealization of a car sill section is shown in Fig. 3.4 and can

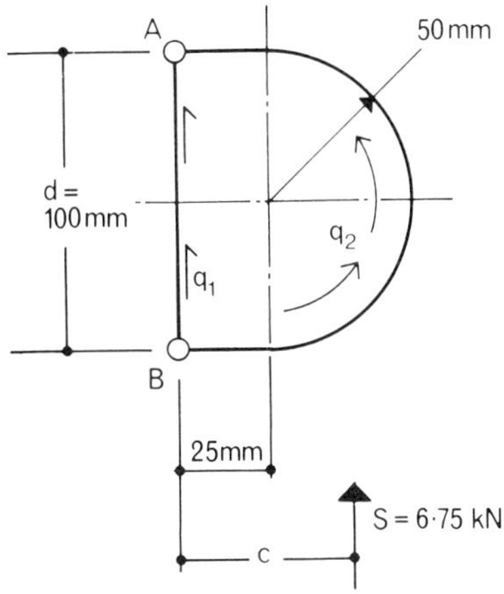

Fig. 3.4. Idealized car sill section

be used to exemplify the above. Note that the section has one axis of symmetry, and since

$$\frac{dP}{dz} = \frac{S}{d} = q$$

in this case $q_1 d + q_2 d = S$. So $q_1 = (S/d) - q_2$ and by taking moments about boom B, $S.c = 2q_2[A]$ thus

$$q_1 = \frac{S}{d} - \frac{Sc}{2[A]} \quad \text{and} \quad q_2 = \frac{Sc}{2[A]}$$

Now

$$A = \frac{\pi \times 50^2}{2} + 100{\cdot}25 = 6{\cdot}43 \times 10^3 \text{ mm}^2$$

so

$$q_2 = \frac{6{\cdot}75 \times 10^3}{12{\cdot}861 \times 10^3} c = 0{\cdot}525c \text{ N/mm}$$

and

$$q_1 = 67{\cdot}5 - 0{\cdot}525c \text{ N/mm}$$

If the vertical web is 3 mm thick and the curved web 1 mm, the position of S for equal shear stress in each web follows from

$$\tau_2 = 3q_2 = 3 \times 0{\cdot}525c$$
$$\tau_1 = 67{\cdot}5 - 0{\cdot}525c$$

so $2{\cdot}05c = 67{\cdot}5$ and $c = 33{\cdot}33$ mm

$$\tau_1 = \tau_2 = 5{\cdot}25 \text{ MN/m}^2$$

EQUATION OF FLEXURE

Sign convention

Most simplified vehicle structural analysis involves an idealization using beam elements; thus a consideration of beams in flexure is fundamental to the study of vehicle structures. Analysis of bending beams is assisted by the establishment of and adherence to a strict sign convention. The element of

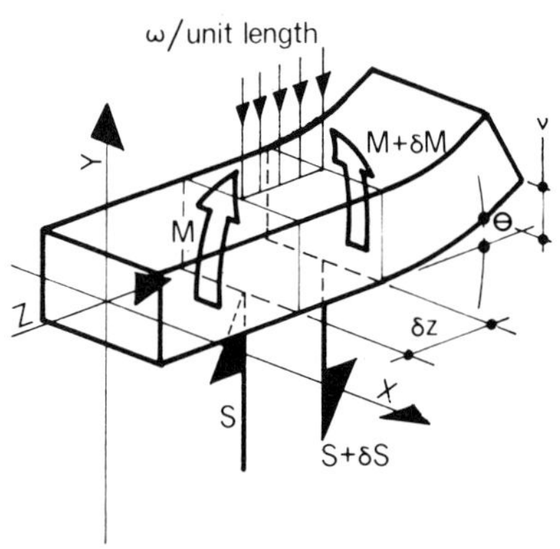

Fig. 3.5. Sign convention for beams in bending

a rectangular-section beam shown in Fig. 3.5 serves to establish the convention and the following four rules will be observed.

1. Shear force S is positive clockwise.
2. Bending moment M is positive upwards towards the compressed top fibres of the beam (moment to the left clockwise; moment to the right anticlockwise).
3. Load is negative applied downwards on to the compressed edge of the beam.
4. Rotational deflection (slope) θ and translational deflection (displacement) v are both positive in the positive direction of upward loading.

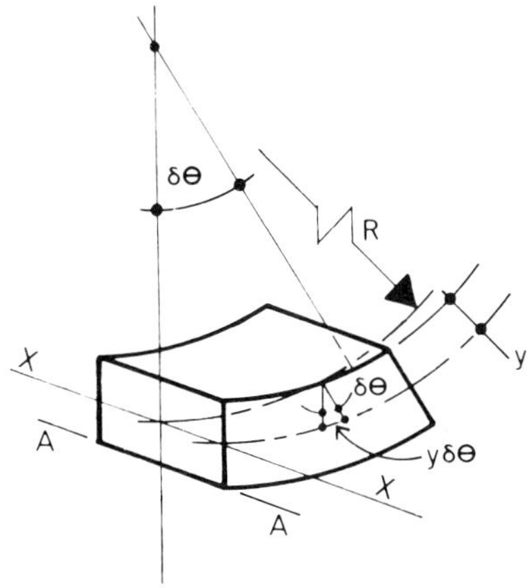

Fig. 3.6. Beam element in bending

Considering the enlarged element of the beam, length δz, in Fig. 3.6; for equilibrium, moments about A–A

$$M + (S + \delta S)\,\mathrm{d}z - w \,.\, \delta z \,.\, \delta z/2 = M + \delta M$$

Multiplying out, cancelling, and discarding terms which are insignificant (multiples of elemental quantities) then

$$S = \frac{\delta M}{\delta z} = \frac{\mathrm{d}M}{\mathrm{d}z}$$

as $z \to 0$.

By geometry $\delta z = R \,.\, \delta\theta$ so $1/R = \delta\theta/\delta z$. From stress/strain relation, at y from the neutral axis

$$\text{Strain } \epsilon = y \,.\, \frac{\delta\theta}{\delta z}$$

so stress $\sigma = Ey/R$. Then moments about X–X through the neutral axis, for unit beam width ($x = 1$)

$$M = \int \sigma \,.\, y \,.\, \mathrm{d}y = \frac{E}{R}\int y^2 \,.\, \mathrm{d}y = \frac{EI}{R}$$

(EI is defined as the flexural stiffness)

$$\theta = \mathrm{d}y/\mathrm{d}z \qquad \text{so} \qquad \mathrm{d}\theta/\mathrm{d}z = 1/R \qquad \text{and} \qquad M/EI = \mathrm{d}^2y/\mathrm{d}z^2$$

Moving along the element in the z direction, shear force increases with $-w$ so

$$w = \frac{-\mathrm{d}S}{\mathrm{d}z} = \frac{-\mathrm{d}^2M}{\mathrm{d}z^2} = -EI\frac{\mathrm{d}^4y}{\mathrm{d}z^4}$$

So

$$\frac{dy}{dz} = \int \frac{M}{EI} \,.\, dz; \qquad y = \int \frac{dy}{dz} \,.\, dz = \iint \frac{M}{EI} \,.\, dz \,.\, dz$$

Referred to as the 'Engineers' Theory' of bending, the above expressions are only valid when plane sections remain plane after bending; this implies that there is uniform shear distribution across the section. The beam is also assumed to be homogeneous, isotropic, initially straight, and to bend in a circular arc of radius of curvature large compared with the cross-section.

Shear in beams

Although shear stress is assumed constant *across* a section it will vary over the depth of the section, as a shear force complementary to the transverse shear due to the external loads. This concept of complementary shear is best understood by examining Fig. 3.7 in which the beam is considered in turn to be composed of vertical and the horizontal slip planes.

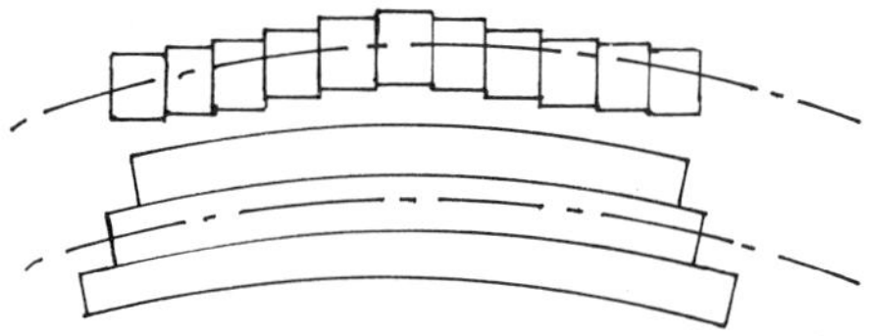

Fig. 3.7. Horizontal and complementary vertical shear

By considering a transverse slice of the beam in Fig. 3.8 and then examining a longitudinal plane cutting the slice at distance y_1 above the neutral axis, there is an area A of section above the cut shown shaded. For an element of the shaded section y above the neutral axis, bending stress $\sigma = My/I$ so the longitudinal force on the element $= \sigma \,.\, \delta A$ and the force over the whole shaded area

$$P = \int_{y_1}^{y_{max}} \frac{M \,.\, y \,.\, dA}{I} \qquad \text{and} \qquad \delta P = \int_{y_1}^{y_{max}} \frac{\delta M \,.\, y \,.\, dA}{I}$$

Shear force $F = \delta P$ and shear stress

$$\tau = \frac{S}{b \,.\, dz} = \frac{\delta M}{b \,.\, dz \,.\, I} \int_{y_1}^{y_{max}} y \,.\, dA$$

But $S = F = dM/dz$ so $\tau = (S \,.\, A \,.\, \bar{y})/(I \,.\, b)$ where y is the distance from the neutral axis to the centroid of shaded area. This is a major expression used in calculating shear. The expression can even be used for finding the

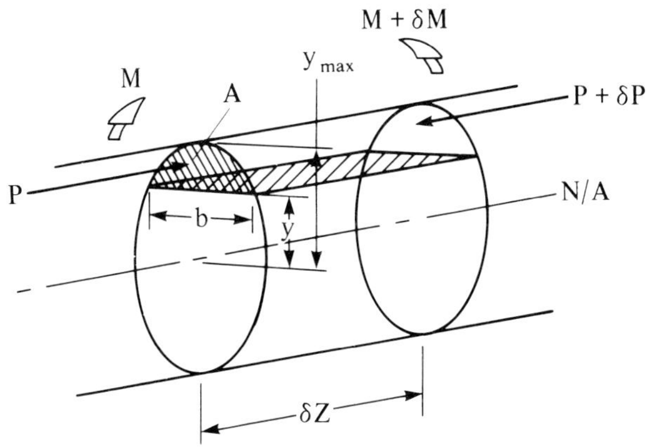

Fig. 3.8. Shear over element of beam section

required pitch of spot welds fastening a reinforcement to a beam. It is written in alternative form as pitch

$$P = \frac{R \cdot I}{S \cdot A \cdot \bar{y}}$$

where R is the spot weld shear load capacity, $\bar{y}$ is distance from the neutral axis to centroid of reinforcement and A the cross-section area of the reinforcement. In the cross section shown in Fig. 3.9 it is required to find the spot weld pitch for attaching the lower reinforcing plate; the section properties can be calculated from Table 3.1 in which I_g is second moment of area of a part section about its own axis.

Distance of neutral axis from lower face of bottom reinforcing plate

$$y_{n.a.} = \frac{\sum A \cdot y}{\sum A} = \frac{2{\cdot}64}{2{\cdot}15} = 1{\cdot}23 \text{ cm}; \qquad I = \sum (I_g + A \cdot h^2) = 3{\cdot}877 \text{ cm}^2$$

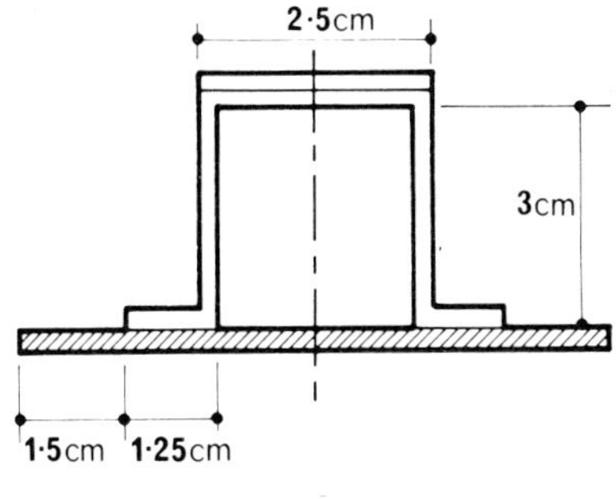

Fig. 3.9. Reinforced top-hat beam section

TABLE 3.1

	Area A (*cm*^{2})	*Centroid distance* y (*cm*)	Ay (*cm*^{3})	I_g (*cm*^{4})	*Centroid distance to N/A,* h (*cm*)	Ah^2 (*cm*^{4})
Top 'flanges'	0·5	3·2	1·60	$(2{\cdot}5 \times 0{\cdot}2^3)/12 = 0{\cdot}00167$	1·97	1·94
Webs	0·6	1·6	0·96	$(0{\cdot}2 \times 3^3)/12 = 0{\cdot}45$	0·37	0·082
Bottom flanges	0·25	0·15	0·04	$(2{\cdot}5 \times 0{\cdot}1^3)/12 = 0{\cdot}00021$	1·08	0·293
Reinf. plate	0·8	0·05	0·04	$(8 \times 0{\cdot}1^3)/12 = 0{\cdot}00067$	1·10	1·11
	$\sum = 2{\cdot}15$		$\sum = 2{\cdot}64$	$\sum = 0{\cdot}4525$		$\sum = 3{\cdot}425$

Distance from neutral axis to extreme fibre $y = 3{\cdot}3 - 1{\cdot}23 = 2{\cdot}07$ cm. Given a bending moment $M = 300$ Nm, bending stress

$$\frac{M \,.\, y}{I} = 300 \times 2{\cdot}07 \times 10^{-2}/3{\cdot}877 \times 10^{-8} = 160 \text{ MN/m}^2$$

Distance from neutral axis to centroid of reinforcing plate

$$\bar{y} = 1{\cdot}18 \times 10^{-2} \text{ m}$$

Given $R = 500$ N for each spot weld and assuming the flanges are held by two rows of spots

$$p = \frac{R \,.\, I}{S \,.\, A \,.\, \bar{y}} = \frac{2{\cdot}500 \times 3{\cdot}878 \times 10^{-8} \times 10^3}{2{\cdot}0 \times 10^3 \times 0{\cdot}8 \times 10^{-4} \times 1{\cdot}18 \times 10^{-2}} = 20{\cdot}5 \text{ mm}$$

'Thin-walled' beams

Thin-walled beams having wall thickness insignificant in relation to the overall section can be shown to have a varying shear stress across a transverse section in contradiction to the assumption of constant shear. Consideration of a thin-walled I-beam as in Fig. 3.10(a) demonstrates the effect.

Within the flange

$$\tau = \frac{S}{I} \,.\, b_1 \int_{y_1}^{d_1/2} y \,.\, b_1 \,.\, \mathrm{d}y = \frac{S}{I}\left\{\frac{d_1^2}{4} - y_1^2\right\}$$

At the inner surface of the flange

$$\tau = \frac{S}{2I}\left(\frac{d_1^2}{4} - \frac{d_2^2}{4}\right) \tag{3.1}$$

Within the web

$$\tau = \frac{S}{Ib_2}\left(\int_{d_2/2}^{d_1/2} y \,.\, b_1 \,.\, \mathrm{d}y + \int_{y_1}^{d_2/2} y \,.\, b_2 \,.\, \mathrm{d}y\right)$$

$$= \frac{S}{I}\left\{\frac{b_1}{b_2}\frac{1}{2}\left(\frac{d_1^2 - d_2^2}{4}\right) + \frac{1}{2}\left(\frac{d_2^2}{4} - y_1^2\right)\right\}$$

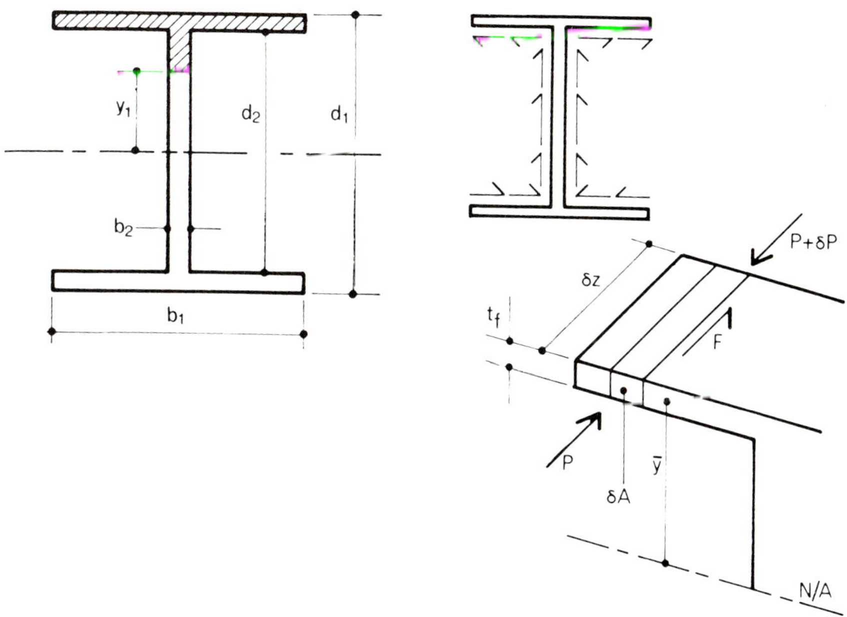

Fig. 3.10. Shear distribution in an I-section beam

At the top surface of the web

$$\tau = \frac{S}{I}\frac{b_1}{b_2}\left(\frac{d_1^2 - d_2^2}{8}\right) \tag{3.2}$$

The discrepancy between equations (3.1) and (3.2) is explained by the constant shear stress assumption. In practice complementary shear flow acts horizontally in the flanges (as well as vertically in the web) running out to zero at the edges, Fig. 3.10(b) and (c). Its average value

$$q_f = \tau \, . \, t_f = \frac{F}{\delta z} = \frac{\delta M \bar{y} \; \delta A}{\delta z \, . \, I} = \frac{S}{I} \, . \, A_f \, . \, \bar{y}$$

So further assumptions are made with thin-walled beams that bending stress is constant over the flange depth and that there is uniform shear vertically on the web.

With certain open-section beams the shear flows around the section can create a torque which can twist the beam unless the external load is applied at an offset just sufficient to balance the couple developed by the shear flows. In Fig. 3.11 the force S acts through the 'shear centre' to satisfy this condition. At distance x from the flange edge

$$q_f = \frac{S}{I}(x \, . \, t)\frac{d}{2}$$

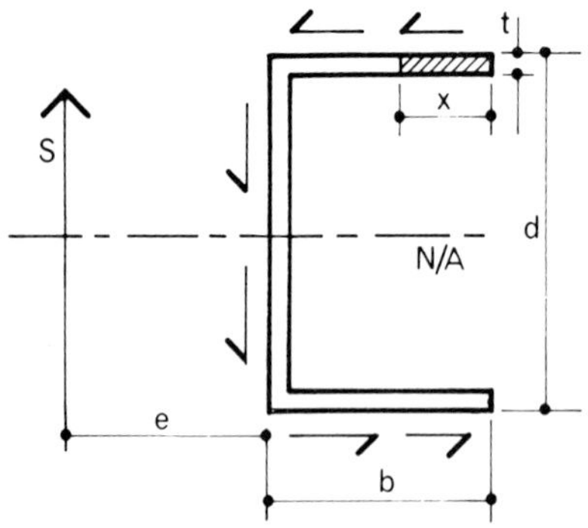

Fig. 3.11. Shear flow in an I-*section beam*

so the shear flow in the flange

$$\int_0^b q_f \,.\, dx = \frac{Stdb^2}{4I}$$

and has a moment arm d. For equilibrium the offset of the shear centre

$$e = \frac{t \,.\, d^2 \,.\, b^2}{4I}$$

An example of an open-section beam is a top-hat section loaded as shown in Fig. 3.12. To find the value of e, the shear centre offset and

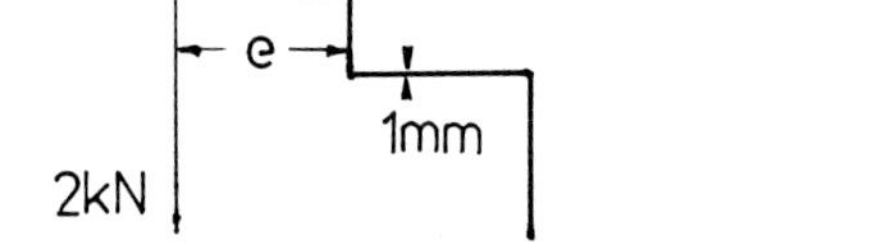

Fig. 3.12. Shear flow in top-hat section beam

expressions for the shear distribution and maximum shear stress for the section, the following procedure is adopted

$$I_{xx} = \frac{0{\cdot}1 \times 8^3}{12} + 2(0{\cdot}1 \times 2{\cdot}2^2) = 5{\cdot}86 \text{ cm}^4$$

$$\frac{S}{I} = \frac{2 \times 10^3}{5{\cdot}86} = 341 \text{ N/cm}^4 \qquad \text{in } q = \frac{S}{I}\int st \, ds + q_0$$

s_1 and s_2 are expressed in terms of y so

$$q_{AB} = 341 \int_4^y y \times 0{\cdot}1(-dy) = 34{\cdot}1\left(\frac{16 - y^2}{2}\right); \; q_B(\text{at } y = 2 \text{ cm})$$

$$= 34{\cdot}1\left(\frac{12}{2}\right) = 204{\cdot}6 \text{ N/cm}$$

$$q_{BC} = 341 \int_0^{s_2} 2 \times 0{\cdot}1 \, ds_2 + 204{\cdot}6 = 68{\cdot}2s_2 + 204{\cdot}6; \; q_C(\text{at } s_2 = 2)$$

$$= 136{\cdot}4 + 204{\cdot}6 = 341 \text{ N/cm}$$

$$q_{CD} = 341 \int_2^y y \times 0{\cdot}1(-dy) + q_C = 34{\cdot}1\left(\frac{4 - y^2}{2}\right) + 341$$

$$= 17{\cdot}05(4 - y^2) + 341$$

$$q_{max} = q_D(\text{at } y = 0) = 17{\cdot}05 \times 4 + 341 = 409{\cdot}5 \text{ N/cm}$$

$$\tau_{max} = \left(\frac{409{\cdot}5}{0{\cdot}1}\right)10^4 \text{ N/mm}^2 = 40{\cdot}95 \text{ MN/m}^2; \text{ then moments about } D\text{:}$$

$$2000 \times e = 2 \int_0^2 q_{BC} \, . \, ds_2 \times 2 - 2 \int_4^2 q_{AB}(-dy) \times 2$$

$$= 4 \int_0^2 (68{\cdot}2s_2 + 204{\cdot}6) \, ds_2 - 4 \int_4^2 17{\cdot}05(16 - y^2)(-dy)$$

$$e = 4(545{\cdot}4 - 208{\cdot}5)/2000 = 0{\cdot}675 \text{ cm}$$

Torsion and twist

To determine shear stress due to twisting of beams and the torsional displacement, a random hollow section as in Fig. 3.13 is considered, being of constant section along its length (b). There is a tendency under torque T for adjacent cross sections to slide over one another, creating shear flow q. Transverse shear force $q \, . \, \delta s$ balances complementary longitudinal shear force F over slice BCDE; thus $F \, . \, \delta s = (q \, . \, \delta s)l$ and $F/l = q$ applies around the section. Moments about any point 0: $\delta s \, . \, 00' = 2 \times$ area $\triangle$OBC. Summing for all the elements gives torque $T = q \, . \, 2[A]$ where $[A]$ is the total area enclosed by the median line of the wall.

Considering the shear of an element in twist (c), then for small displacement $\delta\phi$, twist per unit length $\theta/l = d\phi/dz$ and external work done by torque $T, = T \, . \, d\phi = 2[A]q \, . \, d\phi$ and as $G = q/t\gamma$, internal work done by $q \text{ ds} = q \, . \, ds \, . \, \gamma \, . \, dz = q \, . \, ds \, . \, q \, . \, dz/tG$ which on integrating around the whole section gives

$$\int \frac{q \, . \, ds}{t} \, . \, q \, . \, \frac{dz}{G}$$

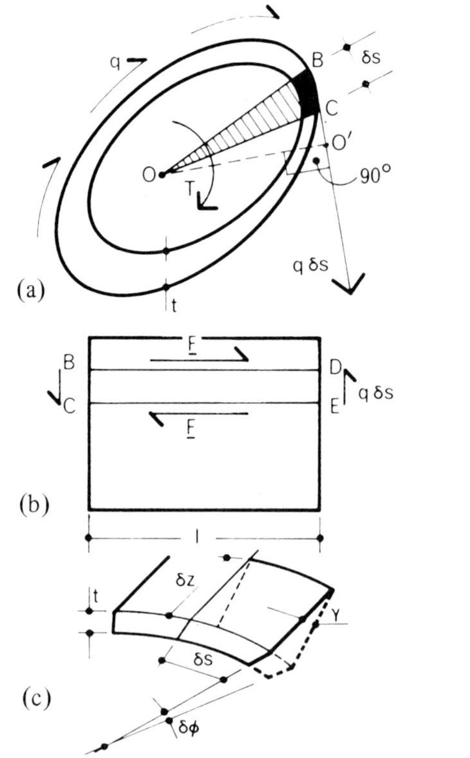

Fig. 3.13. Torsion and twist of a generalized section

So, equating external and internal (virtual) work

$$\theta/l = \frac{q}{2[A]G} \cdot \int \frac{l}{t} \cdot \mathrm{d}s$$

These expressions (the Bredt Batho equations) give shear stress and angle of twist for closed-section members under the action of an external torque; they also give the condition for zero twist (load acting through shear centre) for a laden beam.

The expressions are also reasonably valid for open-section members of very compact section, the limiting condition being that plane sections remain plane and there is no axial warping of the section to induce axial direct stress. For sections made up of equi-thickness rectangles thickness t and width h_1, h_2, h_3, etc.

$$\frac{\theta}{1} = T/G(\tfrac{1}{3}) \propto (\sum h_i)t^3$$

with shape coefficients equal to 0·83 for angles, 1·10 for *C* shapes, 1·17 for zeds, and 1·00 for tees.

Generally for open sections cross-sectional warping is far higher than for closed sections and without warping constraint the torsional stiffness is small. For the thin-walled I-beam under torsion in Fig. 3.14 differential bending of the flanges is taking place, which, when section warping is restrained, accounts for most of the torsional resistance.

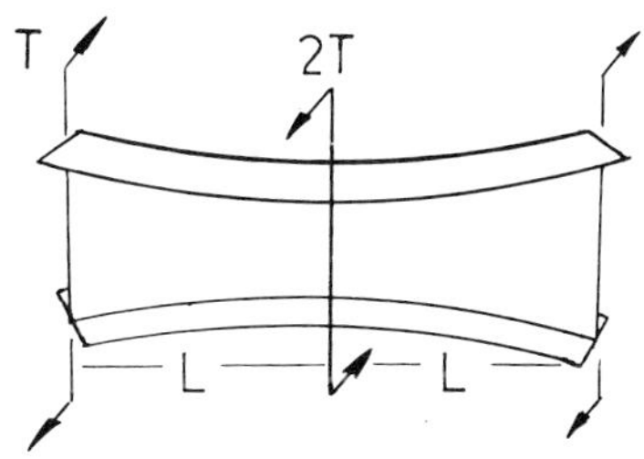

Fig. 3.14. Differential bending in torsion of open-section members

In the design of hollow-section thin-wall beams such as the side members of sports/racing car punt structures it is apparent that the efficiency of the beam will be improved if as much as possible of the cross-section material is placed at the top and bottom, so increasing section second moment of area; but this is with the proviso that the thickness of the web is sufficient to resist buckling under the shear and compressive stresses. Also to exploit strength fully, cross-sectional shape must be maintained, thus creating a need for bulkheads or ring frames.

Beam idealization

Thin-walled beams can be idealized into systems of shear panels and end-load carrying booms, as in Fig. 3.1, whose areas include contributions from the skin panels. A typical symmetrical rectangular-section box beam with angle corner flanges area A_f, depth d, and breadth b with corresponding panel thicknesses t_d and t_b would be idealized by a four-boom box beam with boom section areas

$$A_f + (b \cdot t_b/2) + (d \cdot t_d/6)$$

If web thicknesses are unequal these are taken into account in the Bredt Batho equations and, of course, in calculating panel shear stress. The technique in calculating shear flow is to 'open' the section, removing one element, to obtain 'open section shear flows'. Then an 'opposite-sense shear flow q_0' is introduced on all elements, equal to the 'resultant shear flow', q^*, on the cut element; and final stage is to balance moments of the 'combined shear flows' against the moment arm of the externally applied load. The 'q_0 flows' are assumed to set up torque

$$T = q_0 \,.\, 2[A]$$

from the first Bredt Batho equation, in the moment calculation. When shear centre position has to be found the value of its offset is determined by equating twist per unit length to zero; that is, from the second Bredt Batho equation

$$\sum q^* \,.\, \frac{ds}{t} = 0$$

A typical idealized box-beam section is shown in Fig. 3.15 in which $A = 1\ \text{cm}^2$. Calculation of the shear flow distribution is as follows: Section

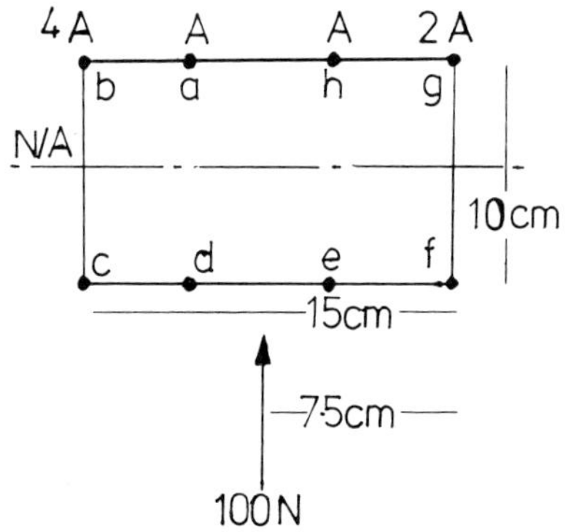

Fig. 3.15. Idealized eight-boom box beam

second moment of area $I = (0{\cdot}8 \times 5^2)2 = 40\ \text{cm}^2$. Now cut *ah* (and by symmetry *de*) then the open-section shear flows, clockwise positive are

$$q_{ab} = \frac{S}{I} \sum A\bar{y} = \frac{100}{40} \sum 0{\cdot}1 \times 5 = -1{\cdot}25\ \text{N/cm} = q_{cd}$$

$$q_{bc} = 2{\cdot}5 \sum 0{\cdot}6 \times 5 = -6{\cdot}25$$

$$q_{hg} = 2{\cdot}5 \sum 0{\cdot}1 \times 5 = +1{\cdot}25 = q_{ef}$$

$$q_{gf} = 2{\cdot}5 \sum 0{\cdot}3 \times 5 = +3{\cdot}75$$

Then superimposing q_0 shear flows in the opposite direction and taking moments about C

$$T = 2[A]q_0 = -100 \times 7{\cdot}5 + 10 \times 3{\cdot}75 \times 15$$

$$- 1{\cdot}25 \times 5 \times 10 + 1{\cdot}25 \times 5 \times 10$$

$$= -187{\cdot}5$$

So

$$q_0 = +187{\cdot}5/2{\cdot}15 \times 10 = +0{\cdot}625lb \quad \text{(clockwise)}$$

$$q^*_{ab} = -1{\cdot}25 + 0{\cdot}625 = -0{\cdot}625 = q_{cd} \quad \text{(anticlockwise)}$$

$$q_{ah} = q_{de} = +0{\cdot}625, \; q_{bc} = -6{\cdot}25 + 0{\cdot}625 = -5{\cdot}63$$

$$q_{hg} = +1{\cdot}25 + 0{\cdot}625 = 1{\cdot}825 = q_{ef}$$

$$q_{gf} = +3{\cdot}75 + 0{\cdot}625 = 4{\cdot}395 \text{ N/cm}$$

Triangular-section beams can be evaluated by the same methods. For example, the section shown in Fig. 3.16 which has equal boom areas, a,

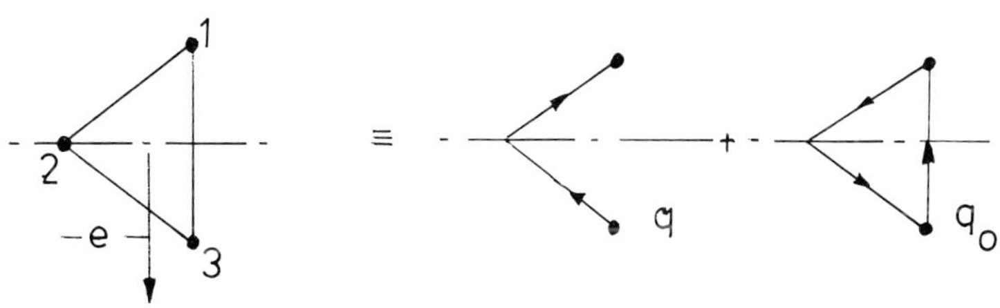

Fig. 3.16. Idealized triangular-section beam

shear panel widths h and thicknesses t_1 and t for vertical and angled panels, respectively. To find the shear centre, the right-hand panel is removed to first obtain open-section shear flows; but by taking moments about (2) q_{12} and q_{23} are isolated and q_0 is balanced by the moment of the external force S at h the shear centre, offset e. By putting $S = I$

$$I = 2a\left(\frac{h}{2}\right)^2 = \frac{ah^2}{2} = S \qquad \text{so} \qquad \frac{a\,.\,h^2}{2} e = q_0 \frac{h\sqrt{3}}{2} h \qquad \text{and} \qquad q_0 = \frac{ae}{\sqrt{3}}$$

Putting $\theta/l = 0$, that is, $\int q^*(\mathrm{d}s/t) = 0$. But open-section shear flows $q_{12} = q_{23} = ah/2$. And

$$q^*_{12} = q^*_{23} = \frac{ah}{2} - \frac{ae}{\sqrt{3}}$$

therefore

$$2\left(\frac{ah}{2} - \frac{ae}{\sqrt{3}}\right)\frac{h}{t} - \frac{Ae}{\sqrt{3}}\frac{h}{t_1} = 0$$

therefore

$$\frac{h}{t} = \frac{2e}{\sqrt{3 \times t}} + \frac{e}{\sqrt{3 \times t_1}} = \frac{e}{\sqrt{3}}\left(\frac{2t_1 + t}{tt_1}\right)$$

therefore

$$e = \frac{\sqrt{3}\,ht_1}{2t_1 + t}$$

Tapered beams can be considered if allowance is made for change in panel shear flow not merely because of the change in depth of the panel but also because some of the external shear force is carried by the vertical components of the flange axial forces. Figure 3.17 shows a panel in which

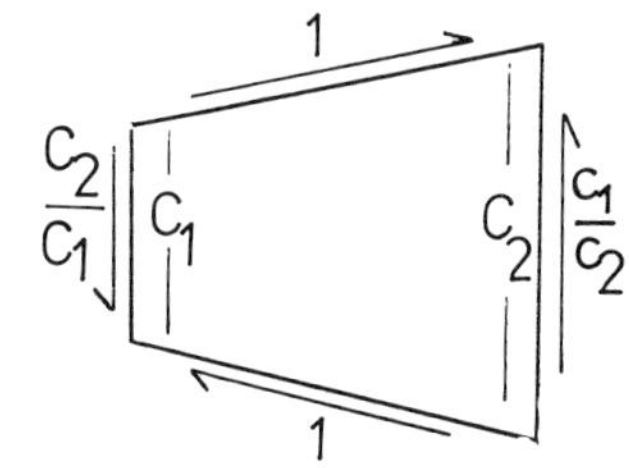

Fig. 3.17. Shear flow in tapered beam

shear flow falls parabolically from c_2/c_1 to c_1/c_2 along its length. For beams which change in section along their length the shear centre also changes; the line which is a locus of the shear centre is known as the flexural axis. In the case of a curved beam, Fig. 3.18, under flexure the web is subjected to direct stress M/hRt where R is the radius of curvature of the

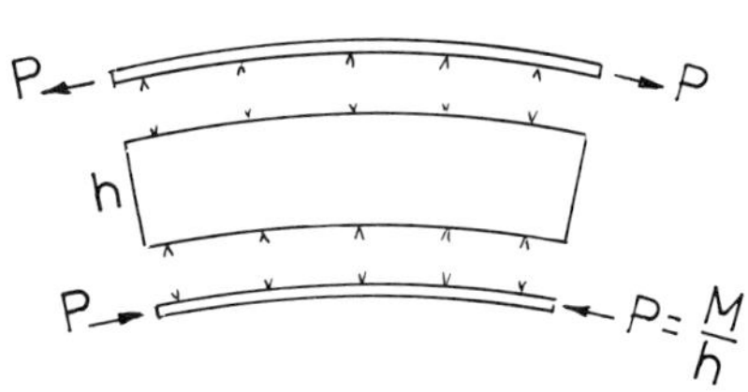

Fig. 3.18. Forces on a curved beam element

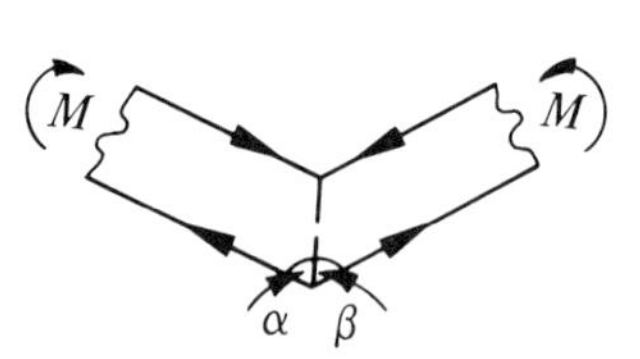

Fig. 3.19. Forces on a kinked beam

neutral axis and t is web thickness. For the kinked beam shown at Fig. 3.19, kink loads are developed at the junction tending to force the flanges to the neutral axis. The tendency should be reacted by inserting a kink strut. Load in the strut will be $M/d(\cos \alpha + \cos \beta)$. Alternatively, beam shear panels may be overlapped so the flange end loads are transferred in shear from the flanges of one into the web of the other; it is usual to taper the flanges where the webs overlap to economize the joint.

INSTABILITY OF THIN-WALLED STRUCTURES

Behaviour of struts

Thin-walled structures more often fail by buckling (local wrinkling or overall 'column buckling') rather than direct overstressing of the material. The instability of struts under load was examined by Euler's classic analysis in which he solved the differential equation of flexure for an end-loaded beam. For the bowed pin-ended strut shown in Fig. 3.20 at distance z along the strut, bending moment $M = Pv$ so that

$$\frac{d^2y}{dz^2} = \frac{-Pv}{EI}$$

where I is the lowest section second moment of area.

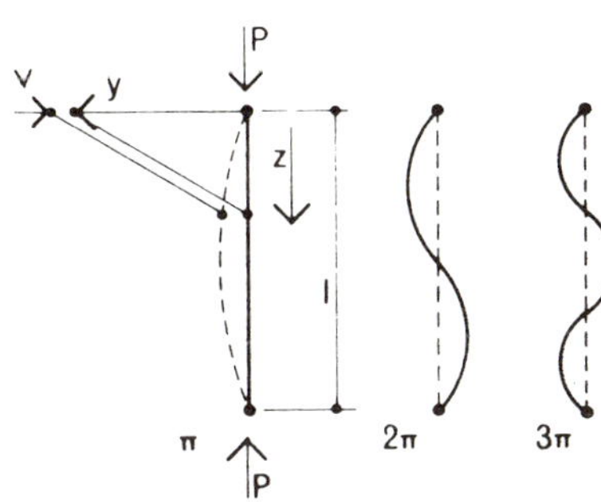

Fig. 3.20. Deflection of an end-loaded strut

Putting $\alpha = \sqrt{(P/EI)}$ in the standard solution

$$v = A \sin \alpha z + B \cos \alpha z$$

Beam end conditions can be used to eliminate A and B so that at $z = 0$, $v = 0$, so $B = 0$ and at $z = 1$, $v = 0$ so $A \sin \alpha l = 0$. If $A = 0$ the strut does not buckle; otherwise $\alpha l = \pi$, 2π, 3π, corresponding to the buckling modes shown in Fig. 3.20. The first is the practical case for which $\alpha^2 = (\pi/l^2) = (P/EI)$ and the Euler critical load

$$P_e = \frac{\pi^2 EI}{l^2}$$

An effective length L may be substituted for l when other than pinned-end conditions exist. L is determined from the points of contraflexure (zero bending moment and directional change of slope) arising with different end conditions; see Fig. 3.21.

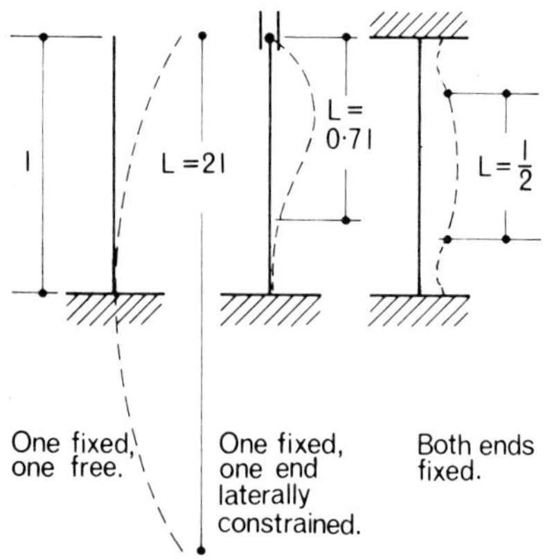

Fig. 3.21. Deflection modes of fixed ended struts

In practical struts a limitation is placed on the Euler theory arising from slight initial curvature and loading eccentricity which exists in real structures. A slenderness ratio is defined $L/K \simeq 120$ (for K = least radius of gyration in $I = AK^2$) below which the Euler equation is inaccurate. In a plot of slenderness ratio against critical stress at which column buckling occurs, three categories of column are revealed. For stocky columns the maximum direct stress is the criterion of failure; intermediate columns require empirical formulae to allow for increased bending moment arising from initial curvature and load eccentricity. An expression due to Rankine fits an 'intermediate' curve tangential to the 'stocky' and 'long' column lines in Fig. 3.22 where

$$\frac{1}{\sigma_{crit}} = \frac{1}{\sigma_{c.max}} + \frac{1}{\sigma_e}$$

from which the critical stress

$$\sigma_{crit} = \frac{\sigma_{c.max}}{1 + (\sigma_{c.max}/\pi^2 E)(l/K)^2}$$

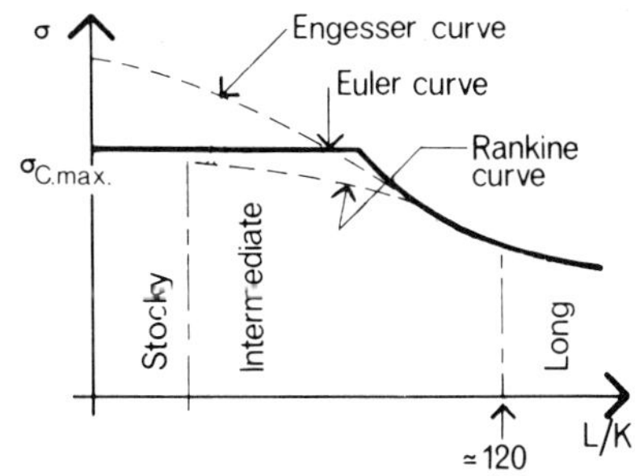

Fig. 3.22. Buckling stress × slenderness ratios for various columns

An example of a somewhat idealized car centre pillar (B-C post) section is shown in Fig. 3.23 to examine for column buckling under a simulated roll-over condition. Determining the neutral axis position and section second moment of area is by the tabular method shown in Table 3.2.

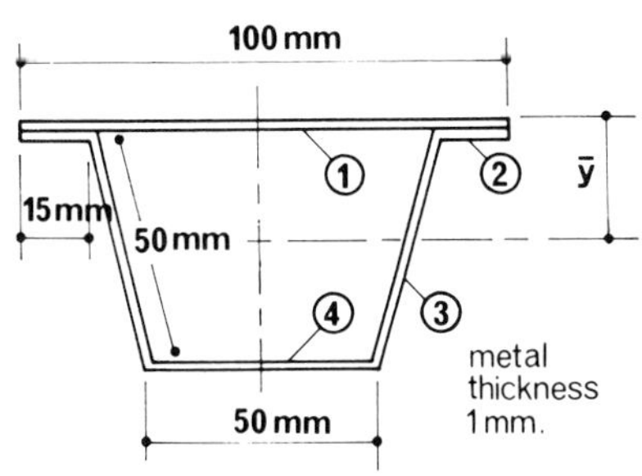

Fig. 3.23. Section through car centre pillar (B-C *post*)

Neglecting any fixing constraint by the roof but assuming the pillar, of one metre length, to be fixed at its base, an effective length $L = 2l$ is used so:

$$P_{\text{crit}} = \frac{10 \times 210 \times 10^9 \times 8{\cdot}4 \times 10^{-8}}{2^2},$$

with $E = 210 \times 10^3$ N/mm^2. Stress at this load $= 44 \times 10^3/280 \times 10^{-6} =$ 157 MN/m^2 since $A = 250$ mm^2. If the B-C post is assumed to have lateral end constraint from the cant rail then $L = 0{\cdot}7l$ and $\sigma_{\text{crit}} = 1280$ MN/m^2 so the strut would not be unstable.

Plate instability

Buckling of rectangular plates under end loading can be examined by extensions of Euler theory. First we consider the Euler buckling load for a

TABLE 3.2

Element	Cross-section area A	Centroidal distance from upper edge of section y	Ay	Section 2nd moment of area of element about an axis through its centroid Ig	Centroidal distance from neutral axis h	Ah^2	$Ig + Ah^2$
1.	100	$\frac{1}{2}$	50	$100/2 = 8{\cdot}3$	18·5	34 200	34 208·3
2.	2×15	$1\frac{1}{2}$	45	$30/12 = 2{\cdot}5$	17·5	9250	9252·5
3.	2×51	$26\frac{1}{2}$	2750	$2{\cdot}50^2/12 = 1040$	7·5	5620	6660
4.	50	51	2550	$50/12 = 4{\cdot}14$	32	34 000	34 004·14
	$\sum = 282$		$\sum = 5395$	$\bar{y} = \sum Ay/\sum A$		$I = \sum =$	84 124·91

narrow rectangular steel strip, but putting $I = bt^3/12$ in $\sigma_{\text{crit}} = (\pi^2 EI)/(Al^2)$ where $A = bt$, then

$$\sigma_{\text{crit}} = \frac{\pi^2 E}{12}\left(\frac{t}{b}\right)^2$$

As the strip is widened to an unsupported rectangular plate under end load the buckling load is factored by $1/(1 - \nu^2)$, where ν is Poissons ratio, to account for anticlastic bending. Then a buckling stress coefficient K is introduced to account for different edge support conditions in

$$\sigma_{\text{crit}} = \frac{K \,.\, \pi^2 E}{12(1 - \nu^2)}\frac{(t)^2}{(b)}$$

Values of K can be obtained from curves such as those in Fig. 3.24 provided by R. W. Peach writing in *Machine Design*, Penton, Dec. 19th, 1963.

ESTIMATING SHEAR STRENGTH—TENSION FIELD BEAMS

The behaviour of a thin-walled panel in shear is such that before the limiting value of shear stress is reached the panel will probably, depending on its size and edge-support conditions, suffer an instability failure. A simplified way is provided by E. F. Bruhn of estimating the onset of shear instability, known as diagonal semi-tension field design (*Analysis and design of flight vehicle structures*, Tristate Offset Co., USA, 1958.)

We consider the *top left* diagram of Fig. 3.25 showing a single bay truss; the action of load P tends to cause the deformation shown *top right*. A parallel situation can be considered for the shear panel, *bottom left* which tends to create, under load P, a stress system shown *bottom right*. Where the compression brace A buckles as a slender column so also does the thin sheet undergo a diagonal wrinkling deformation under stress σ_c. Then as

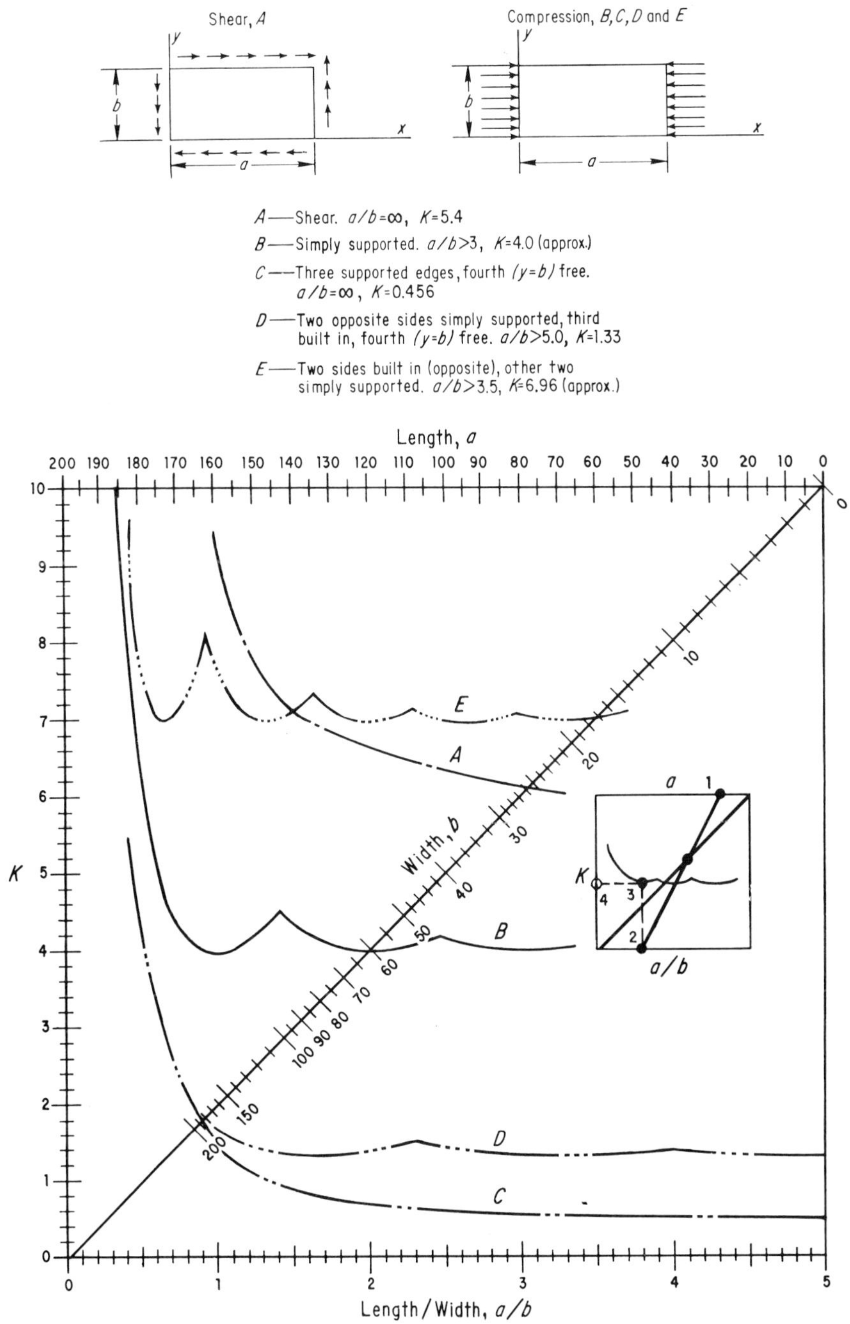

Fig. 3.24. Buckling stress coefficients for various panels

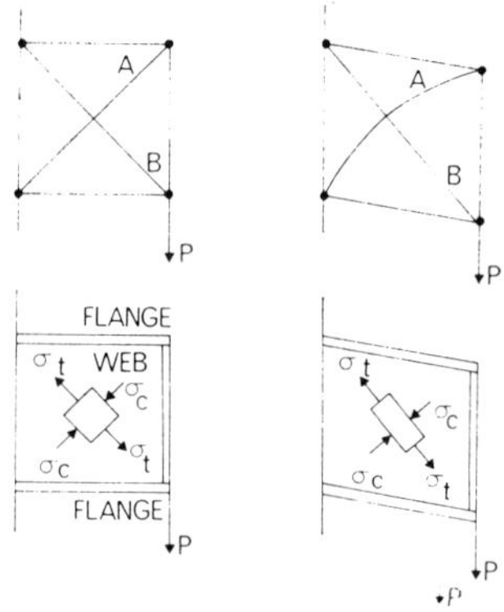

Fig. 3.25. Shear deformation expressed as a tension field

brace B is caused to react higher tension so too is there an increased diagonal tension field in the panel.

In the reinforced shear panel depicted in Fig. 3.26(a), which might represent the side panel of an integral box van, loaded centrally by concentrated force $2S$, wrinkling instability would eventually occur with a limiting value of S. This is depicted by the diagonal broken lines which could also be regarded as tension bands. Assuming that the conditions for a shear panel structural surface exist then the panel can be regarded as in pure shear, τ, and the equilibrium of the shaded triangular element can be represented as in Fig. 3.26(b) from which

$$-\tau \,.\, t \,.\, \mathrm{d}z + \sigma_t \times \frac{t \,.\, \mathrm{d}z}{\sqrt{2}} \times \frac{1}{\sqrt{2}} = 0$$

for panel thickness t and

$$\sigma_v \,.\, t \,.\, \mathrm{d}z - \sigma_t \times \frac{t \,.\, \mathrm{d}z}{\sqrt{2}} \times \frac{1}{\sqrt{2}} = 0$$

whence $\sigma_t = 2\tau$ and $\sigma_v = \tau$.

If the panel is spot welded to the upper and lower flanges at distance apart h between upper and lower rows of welds, then $\tau = S/ht$ and spot welds are subject to vertical and horizontal loading $= \tau \,.\, t \,.\, \mathrm{d}z$ with resultant $= \sqrt{2} \,.\, \tau \,.\, t \,.\, \mathrm{d}z$ so that for $\mathrm{d}z$ length, spot-weld load per unit length $= 1{\cdot}41S/h$. With vertical web stiffeners, holding apart the flanges, each under compressive load P_s and spaced at interval d then

$$P_s = P_1 \sin 45° = \frac{1}{\sqrt{2}} \times \frac{1}{\sqrt{2}} \times \sigma_t \,.\, dt = \sigma_t \times \frac{dt}{2} = S \times d/h$$

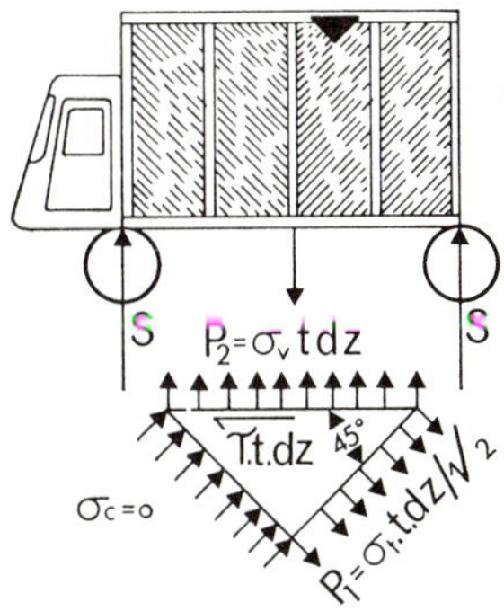

Fig. 3.26. Forces on triangular element of panel in shear

For compressive buckling stress of web stiffeners, channel or zed section, Bruhn provided the expression

$$\sigma_{crit} = K_w \times \frac{\pi^2 . E}{12(1-v^2)} \frac{(t)^2}{(b)}$$

for material elastic modulus E and Poissons ratio v.

The zed or channel section is considered to be a series of plates providing mutual elastic support as in Fig. 3.27. Using the graphs in Fig. 3.28 a buckling stress coefficient k_W is obtained for different relationships of b and t. This expression for critical buckling stress can also be used to determine the axial load capacity of swages by idealizing the swage cross section as a channel with appropriate values of t and b.

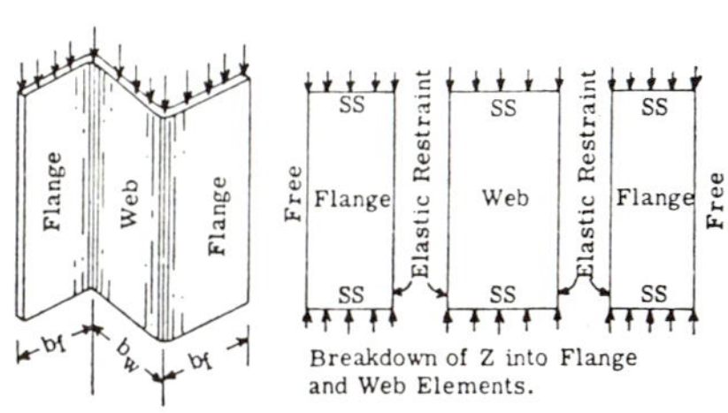

Fig. 3.27. Representation of zed or channel section

PLASTIC BENDING

Examinations of the stress/strain curve for mild steel shows that initial yield occurs at a strain of 0·001 and that the plastic range continues to a strain of 0·010 to 0·020 when strain hardening commences. Because of the material's ductility and long plastic range, increasing the bending moment

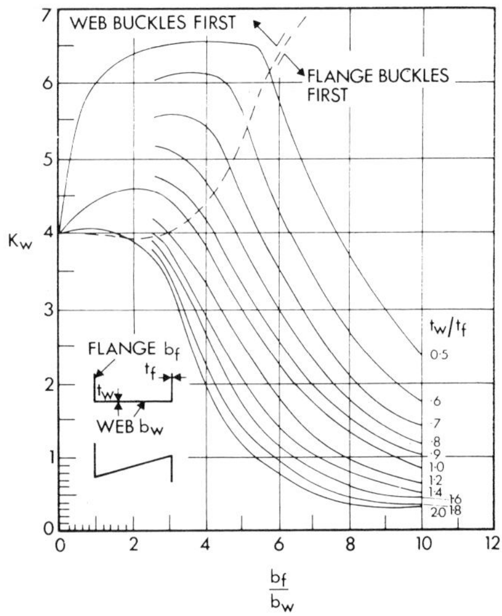

Fig. 3.28. Buckling stress coefficients for channels and zeds

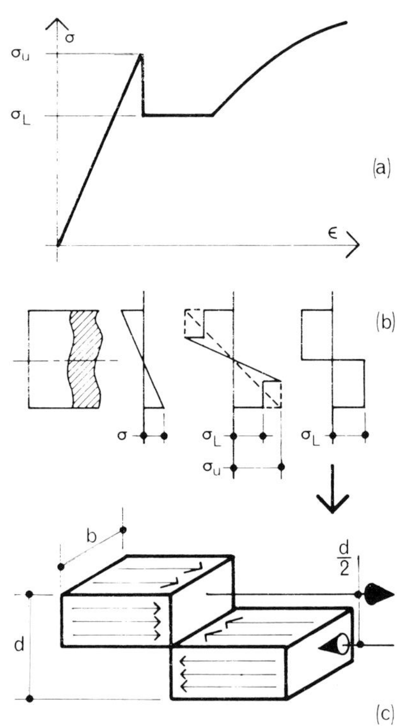

Fig. 3.29. The mechanism of plastic collapse

beyond that at which yield is just reached results in a spread of yielding across the depth of the beam section.

Visualizing an idealized stress/strain curve as in Fig. 3.29(a) and considering the bending stress distribution across a beam section as in (b), the attainment of the upper yield stress σ_U at the extreme fibres is followed by a falling off to the lower yield stress σ_l as bending increases until at failure the lower yield stress applies right across the section. This is defined to be the point for development of a plastic hinge at bending moment M_0 (the plastic moment). From (c) in Fig. 3.29

$$M_0 = \left(\sigma_L \cdot \frac{bd}{2}\right)\frac{d}{2} = \sigma_L \cdot \frac{bd^2}{4}$$

But Engineer's Theory of bending in the elastic range gives

$$M_y = \sigma_y \cdot \frac{bd^2}{6}$$

M_0/M_y $(= 1{\cdot}5)$ for a rectangular section beam is defined to be the 'form factor' for the beam giving a measure of the additional load which can be carried under part plastic strain. For an I beam, form factor is 1·15 and for a circular section tube 1·7.

BENDING OF THIN-WALLED TUBE TANK BODIES

Probably the most common chassisless vehicle structure first to be seen on public roads was the monocoque tanker semi-trailer. The long cylindrical shells formed efficient load carrying beams of originally circular section. The requirement for higher capacity tanks within the same silhouette as the straight circular section ones led to elliptical, and so called 'max-section', units cranked along their lengths as in Fig. 3.30. The transport marketing departments of the major aluminium companies have been prominent in

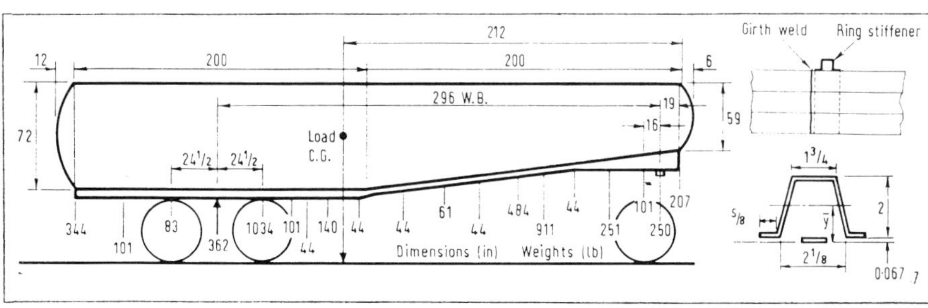

16
17

Fig. 3.30. Frameless tanker trailer structure

developing semi empirical methods for tanker analysis; the following approach is typical: large thin-walled shell beams are more likely to become unstable under extremes of structural loading than to reach limiting stresses in direct tension, compression, or shear. For the compressive edge of the laden circular tank shown in Fig. 3.30, critical stress, found from simple beam theory: $\sigma = My/I$, for the onset of buckling, is given by the empirical expression $\sigma_{crit} = 0{\cdot}38Et/r$.

In Fig. 3.31 the shear force diagram is drawn for the beam, stabilized by stiff rings at pitch L, and the critical shear stress is given by

$$\tau_{crit} = 0{\cdot}313E\beta/(r/t)^{3/2}$$

where

$$\beta = 1{\cdot}2\{(2r/t)^{1/2}/(L/2r)\}^{1/2}$$

for shell thickness t and radius of curvature r. The shear stress maximum is found from the shear force diagram and the section properties.

With the tank supported on the rear bulkhead and adjacent stiffener ring, the bending moment in the ring is given by

$$M_s = 0{\cdot}015\rho D^3 L$$

for a product density ρ and tank ring diameter D. In calculating maximum bending stress from M_s it is usual to assume a $30t$ width of the tank acts with the stiffener in obtaining I_s in

$$\sigma_s = M_s \,.\, y_s/I_s$$

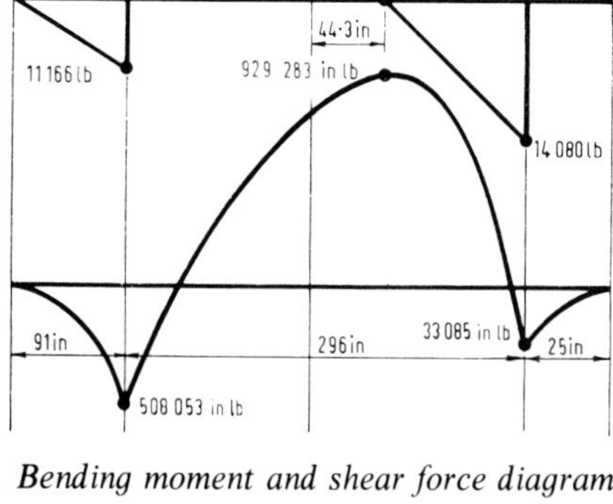

Fig. 3.31. Bending moment and shear force diagrams for tank

For the example in Fig. 3.30, a circular section steel tank reinforced by stiffener rings at $L = 21$ in (0·534 m) pitch, by taking moments about the centre of gravity the load reactions at A and B are

$$R_A = 162\,00 \text{ lbf (77 kN)} \qquad \text{and} \qquad R_B = 305\,15 \text{ lbf (135·8 kN)}$$

Vertical loading at each ring was found to be as in Fig. 3.30 based on 41 850 lbf (186 kN) payload split 55/45 in rear/front shells. Maximum bending moment from Fig. 3.31 was found to be 44·3 in (1·127 m) from the rear/front shell transition, with shell diameter at this point = 69 in (1·75 m). Critical bending stress at this point $\sigma_{crit} = 0{\cdot}38ET/r =$ 22 000 lbf/in^2 (151·5 MN/m^2) for $E = 30 \times 10^6$ lbf/in^2 (206·5 GN/m^2), $t = 0{\cdot}067$ in (1·7 mm) and $r = 34{\cdot}5$ in (0·8775 m). But the maximum bending stress $\sigma_{max} = My/I = 929\,283 \times 2/250 = 7434$ lbf/in^2 (51·3 MN/m^2) giving a safety factor $\simeq 3$. At the point of maximum shear force, critical shear stress

$$\tau_{crit} = \frac{0{\cdot}313E}{(R/t)^{3/2}} \qquad \text{for } \beta = \sqrt{\left\{\frac{(2r/t)^{1/2}}{L2r}\right\}}$$

so $\tau_{crit} = 9660$ lbf/in^2 (66·6 MN/m^2) whereas the maximum shear stress $\tau_{max} = 4800$ lbf/in^2 (38·1 MN/m^2) giving a safety factor of $\simeq 2$ in shear.

For the reinforcing rings, section moment of area

$$I_{\bar{y}} = I_0 + Ax^2 = 0{\cdot}4295 \qquad \text{for 0·09 in (2·28 mm) thick ring material.}$$

So for a 30 in (0·763 m) radius ring, section modulus = 0·413

$$M_s = 0{\cdot}015\rho D^3 L = 5360 \text{ in lb} \qquad \text{for } \rho = 6{\cdot}6/231 \text{ lb/in}^2$$

and

$$\sigma_s = 5360/0{\cdot}413 = 13\,000 \text{ lbf/in}^2 \text{ (8935 MN/m}^2\text{)}$$

maximum bending stress in ring.

CHAPTER 4

IDEALIZATION FOR ANALYSIS: THE UNIT LOAD METHOD AND STRUCTURAL DEFLECTIONS

VEHICLE structures are complex in shape, often because of the constraints imposed by stylist and conceptual designer. It is necessary to idealize them into simplified structures which can be readily analysed. This is so that the load paths through the structure can be visualized as early as possible in the design programme. A representative system of members is needed which can be analysed for overall structural deflection as well as to show internal load magnitude and direction.

A way of idealizing beam elements into end-load carrying bars and shear panels has been introduced in Chapter 3. This is a useful idealization for many of the structural panel sub-assemblies of the vehicle. Other assemblies such as car saloon sideframes which comprise mainly pillars and rails can be idealized as a series of beams in bending with fixed joints.

At the clay model stage of the design programme for a car the design analyst can consider: contour of the skin panels for buckling rigidity, location and types of joints, number and size of roof pillars, size and shape of sill, boot and bonnet counter-balancing forces, and bumper resistance.

Flatness of the skin panels determines susceptibility to buckling; presence of joggles gives warning of joint weakness; configuration of pillars affects the structural contribution of the roof; the sill, being a key member in integral construction, must not be unduly compromised by styling considerations. The development of a structural 'feel', through the understanding of load paths, allows the designer in time to make confident decisions.

MAIN STRUCTURAL ELEMENTS OF A SALOON CAR

A number of different load cases on the body structure must be considered in addition to the road load inputs discussed in Chapter 1. Examples are: resistance to tie-down loads occurring in the rail shipment of cars, adequate anchorage for seat belts, bumper impact resistance, overloading of opened doors by misuse, and jacking at extreme front or rear. Important sections which contribute to overall structural stiffness are roof rails, windshield and backlight headers and pillars, door hinges, front and rear underbody rails, rocker panels, and hinge pillars.

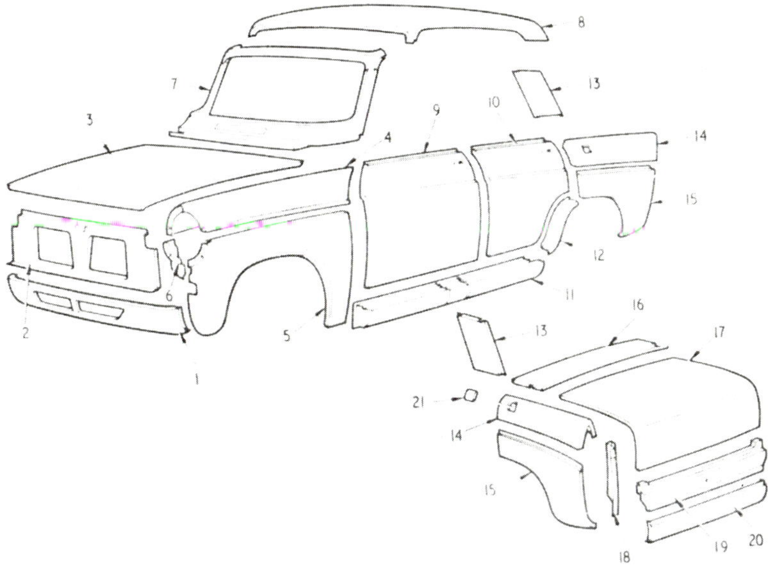

Fig. 4.1. (a) Skin panels nomenclature. (1) body panel front lower, (2) grille panel, (3) bonnet panel, (4) front wing upper, (5) front wing lower, (6) headlamp filler, (7) windshield panel, (8) roof panel, (9) front door, (10) rear door, (11) sill panel, (12) lower 'D' post, (13) rear quarter panel, (14) tonneau panel upper, (15) tonneau panel lower, (16) rear decking, (17) trunk lid, (18) tonneau filler panel, (19) body panel rear upper, (20) body panel rear lower, (21) petrol filler box lid

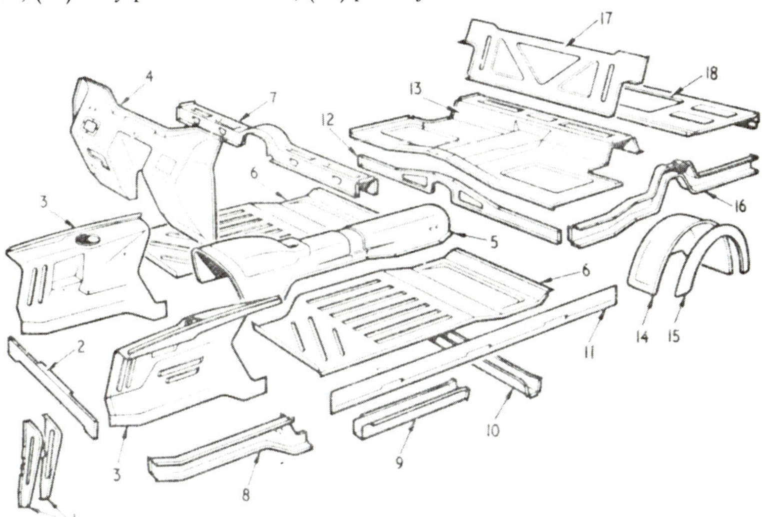

Fig. 4.1. (b) Underbody panels nomenclature. (1) suspension mounting reinforcement, (2) front crossmember, (3) valence panel, (4) dash panel, (5) transmission tunnel, (6) front floor, (7) front floor reinforcement, (8) front longitudinal, (9) extension front longitudinal, (10) front floor crossmember, (11) inner sill panel, (12) rear seat heel-board panel, (13) rear seat pan, (14) rear wheelarch inner, (15) rear wheelarch outer, (16) rear longitudinal, (17) squab panel, (18) luggage floor panel

All the above terms for car body members are American in origin but are gradually becoming accepted usage in the UK. Traditional UK terminology is given in Fig. 4.1 for skin panels and underbody; sideframe nomenclature is defined in the Glossary. The above list would convert to cantrails, windshield header panel (same), backlight reinforcement, door hinges (same), front and rear longitudinals, sills, and tonneau panels, A post and B–C post. Bonnets become hoods; boots become trunks and so on; also changes in construction such as the recent introduction of the one-piece side panel will bring further changes in nomenclature. It will be useful at this stage to describe the traditional construction of a typical integral car body structure as a basis for understanding the part played by different structural elements. It must be appreciated that the development of multi-welders, semi-automated handling methods and such like will have superseded the methods described below and some form of 'gate-line' will now be more common. However, the literature describing such methods is particularly sparse.

The first stage is welding up of the underframe members on jigging which ensures accurate location of suspension anchorages and power unit mounting. According to the configuration of the box-section longitudinals, front and rear sections of the body may be formed either in unit with the underframe or as separate sections which are subsequently welded on. The completed underframe assembly will usually, therefore, consist of front longitudinals with power unit and suspension attachments, side members, body floor, rear-seat pan, and trunk floor or spare wheel housing. Rear suspension attachments are welded to the underside of the rear portion.

For the outer body shell assembly usual practice is to start with left- and right-hand complete sideframes, which will themselves be pre-fabricated. A conventional sideframe assembly consists of a one-piece pressing incorporating A and D posts, the rocker panel, door centre pillar, and cant rails. To this pressing, dash side reinforcements, sill reinforcements, and rear-wheel arch inner panels are spot welded. Initial joining of the right- and left-hand side panel assemblies is at the rear by the trunk panel.

Then follows the assembly of the roof panel; usual practice is to form in one piece the roof pressing, the top of the outer dash, screen pillars, and rear quarters. This one piece pressing runs from the dash sides to the trunk panel. Finally, the whole monocoque shell, less doors and fittings, is formed by welding together the outer body shell assembly to the underframe.

OBSERVED DEFLECTIONS OF CAR STRUCTURES

The way in which the car body structure reacts the principal input loads has been examined in a paper by C. J. Cooke read at the Institution of Mechanical Engineers' one day seminar on 'Vehicle structural analysis' on

15 May 1968. He found, for example, that the significant factor governing passenger compartment torsional stiffness is equal and opposite transverse shear forces in the roof and floor, their development being influenced substantially by the fit of the windscreen and backlight, Fig. 4.2. He also considered the contribution made to the torsional stiffness (by the sills acting as a pair of torque tubes) forming the sides of a peripheral frame, which would include the front dash/bulkhead and the rear seat pan/bulkhead. It is likely that the sills have greater significance in 'open' rather than the 'closed' car structures considered by Cooke.

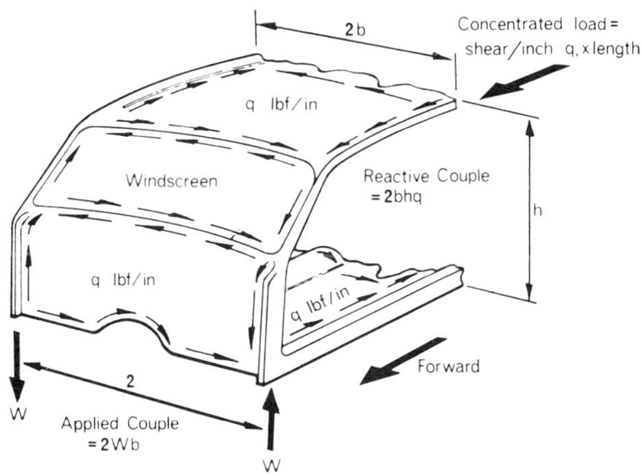

Fig. 4.2. Shear flow over saloon car body in torsion

There was flexing of the forward extensions from the bulkhead arising from the fact that principal axis of the beams formed by the wing and valence assemblies were inclined to the vertical; see Fig. 4.3. The result was that there was a component of the applied force bending the beams about their weak axes to an extent that the radiator frame was unable to stabilize the assembly. Only good continuity between wing and underbody side rail extension, via the bulkhead, ensured that the lozenging displacement of the lower A-post had a positive stiffening effect on the assembly.

Another aspect of 'overall wheelbase' flexure lies in the relative offset from the passenger compartment bulkheads of the front and rear axles; larger front offset normally involves a greater front-end couple in the symmetrical bending mode, and causes a 'lozenging' of the sideframes in which the roof is thrust backwards to produce a larger couple at the D-post which may even result in reverse bending of the rear-end extension, Fig. 4.4.

Cooke pointed out that a surprisingly large improvement in bending stiffness is effected if the screen pillar has a high angle rake in side-elevation and is rigidly jointed to the cant rail. Thereafter care is needed in assessing

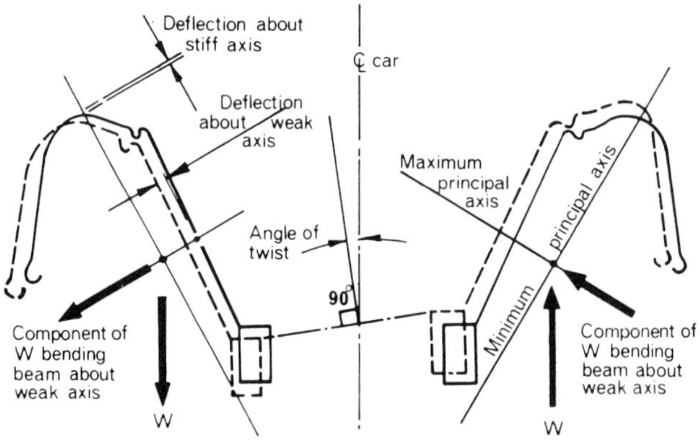

Fig. 4.3. Deflections of wing, valence, and longerons in car front end structure

the effects of the adjacent roof panel to determine an 'effective' cantrail section in the analysis, particularly in relation to allowances for the fore and aft curvature of the roof panel.

Contributors to torsional stiffness

In a more specific paper, (*Proc. Inst. Mech. Engrs.* 1965–66 **180**, part 2A, no. 1) Cooke defined the two main contributory assemblies affecting torsional stiffness as the open-box tube of the passenger compartment and the floor frame, providing expressions for the stiffness of each assembly. He then isolated the additional effect of the B–C post, Fig. 4.5. Displacement of the centre-pillar top end for 1 degree whole body twist is $= 2\pi bh/360a$ for $a = \frac{1}{2}$ length of sill and $b = \frac{1}{2}$ width of passenger compartment so

$$H_{BC} = 3\,\frac{EI_{BC}}{h^2}\left(\frac{2\pi b}{360a} - \theta_{BC}\right)$$

In the underbody cross member between dash and rear squab bulkheads, a torsional resistance is set up according to the difference in slopes of the sills at either end of the member, 2ϕ in Fig. 4.6. Resistance forces are

$$V = T_{cr}/2a = \phi(J_{cr}\,G/2ab)$$

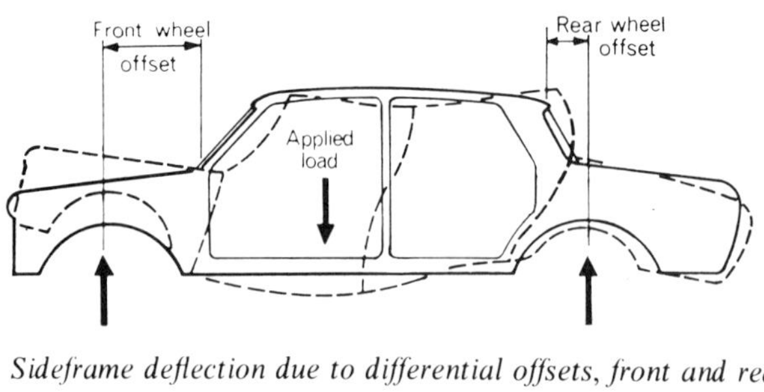

Fig. 4.4. Sideframe deflection due to differential offsets, front and rear wheels

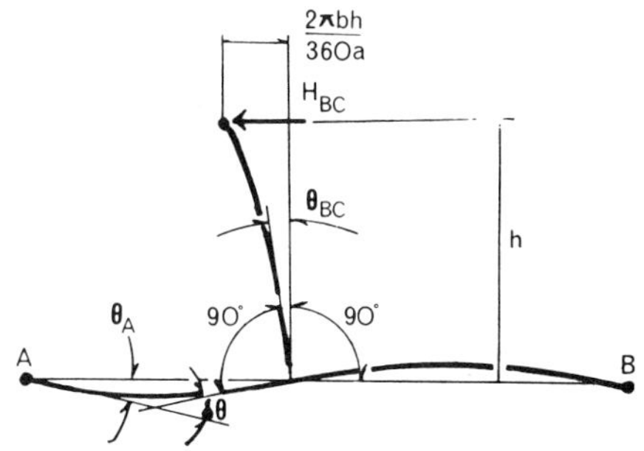

Fig. 4.5. B–C post and sill bending parameters

where

$$2\phi = \frac{\pi b}{360a}$$

and J_{cr} is polar moment of inertia of cross member assuming that the sills are comparatively rigid in bending.

Using these methods a figure of 23 600 lbf ft/deg (32 kN . m/deg) torsional stiffness was obtained for the Rolls Royce Silver Cloud body against 18 300 lbf/deg (24·77 kN . m/deg) measured. Sectional properties for the main pillar members were

$$I_{A\text{-post}} = 8{\cdot}85 \text{ in}^4 \text{ (368 cm}^4) \qquad J_{A\text{-post}} = 16{\cdot}3 \text{ in}^4 \text{ (678 cm}^4)$$

$$I_{B\text{-post}} = 20{\cdot}0 \text{ in}^4 \text{ (833 cm}^4) \qquad J_{B\text{-post}} = 0{\cdot}5 \text{ in}^4 \text{ (20·8 cm}^4)$$

$$I_{wsp} = 0{\cdot}125 \text{ in}^4 \text{ (5·2 cm}^4) \qquad I_{D\text{-post}} = 2{\cdot}0 \text{ in}^4 \text{ (83·2 cm}^4)$$

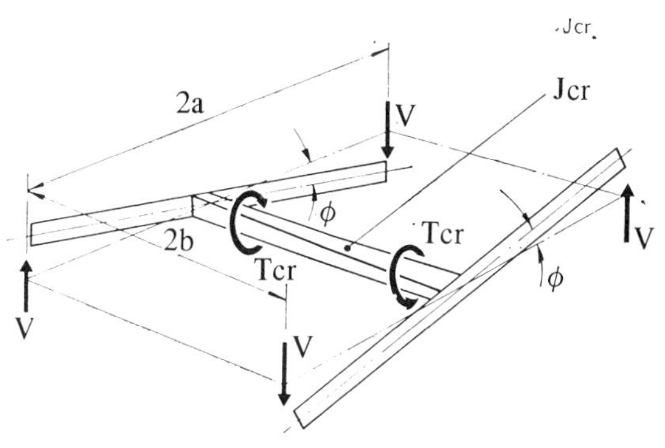

Fig. 4.6. Sill and cross member twist parameters

LOAD LEVELS IN A SALOON CAR STRUCTURE

T. K. Garrett in his article 'Structural Design' *Automobile Engineer*, March, 1953, has examined a saloon car structure under the load systems described in Chapter 1. For obtaining shear force and bending moment diagrams for the vehicle in symmetrical bending he considered the laden vehicle under vertical acceleration of $4{\cdot}5g$ ($3g$ dynamic load effect × 1·5 safety factor).

The structure and load distribution can be idealized as a beam under point loads representing luggage, passengers, power unit, and a uniformly distributed load representing the sprung weight of the vehicle:

$$\frac{4{\cdot}5(135 - 20 - 25)}{3{\cdot}25} = 125 \text{ kg/m} \, (84 \text{ lb/ft})$$

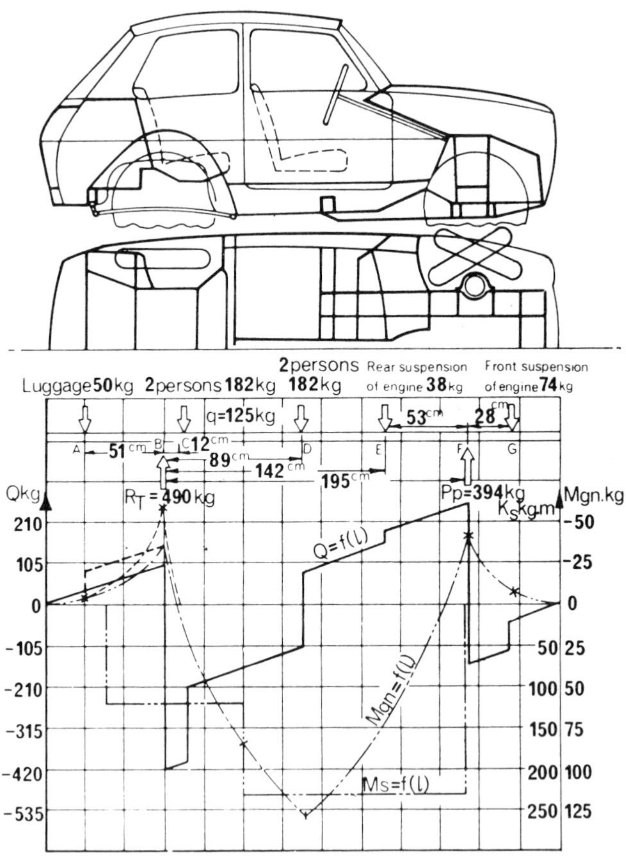

Fig. 4.7. Shear force and bending moment diagram for small saloon car body

This applies to a vehicle with kerb weight 297 lb (135 kg) unsprung weight 44 lb (20 kg), power unit 55·12 lb (25 kg) and vehicle length 10·65 ft (3·25 m).

The simplified structure is shown in Fig. 4.7 with loading diagram, shear force (Q kg) and bending moment (M kgm) diagrams. Only the heavy outlined part of the structure is considered to react the load, the roof being discarded for the purposes of symmetrical bending. The analysis thus only has real significance in convertible-type cars or those with non-structural roofs, since sideframe analysis shows sill strength to be only partially contributory to the overall structure.

For this vehicle, a 5 in (127 mm) deep sill section was analysed using maximum value from the diagrams, of 1900 lb (8·45 kN) shear force and 51 000 lbf in (5·76 kN m) bending moment. For sill section second moment of area 1·348 in^4 (56 cm^4), maximum bending stress was 20·7 tonf/in^2 (32 MN/m^2). Rear quarter and sill junction is one of the most critical joints; in this example the slenderest section of the inner quarter panel, just behind the D post, Fig. 4.8, is 5 in (127 mm) deep. The bending moment corresponding to that section induced tensile and compressive end loads of 33 000 lbf (146·7 kN) for which 12 tonf/in^2 (18·5 MN/m^2) safe working stress of mild steel dictated an area of 0·123 in^2 (0·795 cm^2) of edge material, some 3·4 in (86·5 mm) of 20 swg sheet.

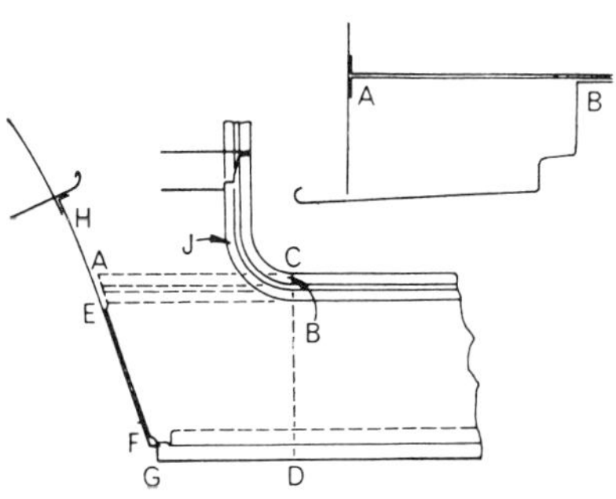

Fig. 4.8. Sill and rear quarter panel junction

Proper precaution must be taken against buckling for which the critical area is the junction of the inner quarter panel and heel board at A. In such a tee connection material is effective against buckling at a distance 16 × metal thickness from A along the three arms of the tee and the flanges of the connection. This area for 20 swg material amounts to 4 × 0·575 = 2·3 in (58·5 mm) length of sheet. The deficiency of material is overcome by extending the sill panel upwards and inner quarter panel downwards so that they overlap and double the strength of the web portion.

Stiffness effects combined

Both symmetrical (bending) and asymmetrical (torsion) cases have to be considered corresponding to the vehicle striking double bump (raising both front wheels) or single bump (raising a single front wheel). It has been suggested that a torsional strength criterion should be the ability of a vehicle to withstand static wheel deflection with single wheel resting on the following bump heights according to vehicle category. Private car 0·788 in (20 cm). Truck 1·18 in (30 cm). Bus 0·985 in (25 cm). Cross country vehicle 1·970 in (50 cm). For K_1 and K_2 overall roll stiffness of the front and rear suspension, the structure assumed to be infinitely stiff, and for track b and bump height h the resulting applied torque

$$T = \frac{K_1 \, . \, K_2}{K_1 + K_2} \cdot \frac{h}{b}$$

while remaining wheels stay in contact with the ground.

Or if the torsional stiffness K_3 of the structure is low

$$T = \frac{K_1 \, . \, K_2 \, . \, K_3}{K_1 + K_2 + K_3} \cdot \frac{h}{b}$$

$$= 1/(f_1 + f_2 + f_3)$$

putting flexibility $f = 1/K$. This technique of adding flexibilities is also useful for determining the overall body stiffness from that of particular structural sections.

CAR BODY IDEALIZATION

J. Pawlowski has described the load paths through a car structure (*Vehicle Body Engineering* Business Books, (1969)). The first stage in the idealization of a whole body, he suggests, is a division into structural surfaces on which the edge loads can be calculated. The three basic system categories of structure can be considered: closed integral, open integral, and punt structure.

The simplest closed box idealization Fig. 4.9(a) serves to show the necessity for vertical panels (shown shaded) to carry floor loads out to the sills. So force P in symmetrical bending is balanced by shear forces S_1, S_2, and in turn by reaction R. The roof and floor are unloaded in shear for this mode and their contribution to bending resistance is restricted to boom area additions. Figures 4.9(b)–(d) shows other simplified idealizations for the three categories of car structure. By contrast, in the torsion case, shown in Fig. 4.10 roof and floor make a major contribution in shear force reaction.

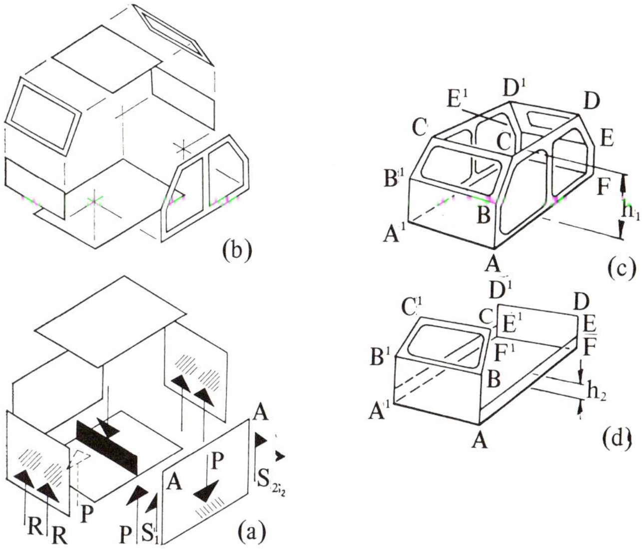

Fig. 4.9. Elements of idealized closed integral, open-integral, and punt type structures

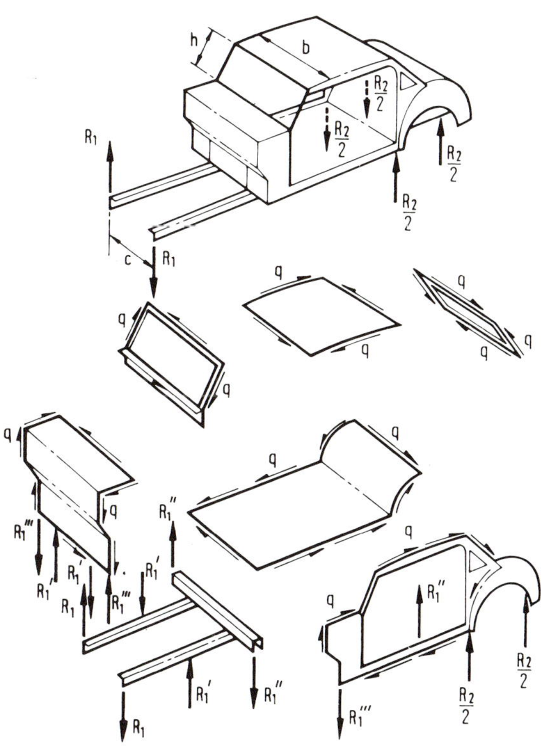

Fig. 4.10. External loads and shear flows on idealized saloon car body in torsion

It is valuable as a next stage to carry out an idealization of a statically determinate set of shear panel and bar representations of the structure; this involves expressing the edge loads on the structural surfaces as shear and end loads in the panels and bars which are the minimum number to support the load system.

A punt structure with built-up bulkheads is considered in the following example as typical of an open-sports or GT car body construction. The structure in Fig. 4.11 (i) is built up from a basic punt shown at (ii). This element is subject to cantilever reactions, under a central load, which give rise to torsional shear and bending loads on the front cross member, in turn reacted in the side members.

Addition of front and rear bulkheads to take suspension loads improves the structural efficiency. The loading introduced through the front cantilever is shown at (iii) and at (iv) a and b are the overall edge load systems in bending and torsion, respectively.

Reactions at the bulkhead due to wheel input loads are:

Vertical	$Z = R_{fp}$ and $X_z = R_{fp} \cdot a/h_1$
Longitudinal	$X_{xg} = R_{fx} \cdot h_1''/(h_1' + h_1'')$ and $X_{xd} = R_{fx} \cdot (h_1' + h_1'')$
Lateral	$Y_g = R_{fy} \cdot h_1''/(h_1' + h_1'')$ and $Y_d = R_{fy} \cdot h_1'/(h_1' + h_1'')$
Longitudinal	$X_{yg} = R_{fg}(a/b)h_1''/(h_1' + h_1'')$ and $X_{yd} = R_{fg} \cdot (a/b)h_1'(h_1' + h_1'')$

The lower (subscript d) forces normal to the front bulkhead can be transferred through the floor structural surface. Upper (subscript y) forces require at least three additional structural surfaces to those shown in (iv) to react the loads.

SYMMETRICAL BENDING

Here the main reaction forces are

$$K_1 = R_{fp},\ K_3 = X_1 = R_{fp}(a_1/h_1),\ K_2 = R_{rp},\ K_4 = X_2 = R_{rp}(a_2/h_2)$$

and

$$K_5 = X_1 - X_2 = R_{fp}(a_1/h_1) - R_{rp}(a_2/h_2)$$

K_3, K_4, and K_5 forces: K_3 in *b–c* causes $q_2 = K_3/l_1$ in panel 7 and its adjacent bars; a balance of forces on bar *b–g* gives downward force at *g* of $P_1 = q_2 h_1$ in turn causing a shear flow in panel 8 of $q_3 = p_1/h$. Flows q_2 and q_3 sum to give $P_2 = K_3(1 + h_1'/h)$ in *g–h*. To find q_4 in panel 9 there is a force balance on *h–m* of $P_1 - q_3 h + q_4 h = 0$ giving $q_4 = (-q_3 h - q_2 h_1)/h = 0$.

In panel 12 $q_6 = k_4/l_4$; on bar *N–J* $P_3 - q_4 h - q_5 h = 0$ so

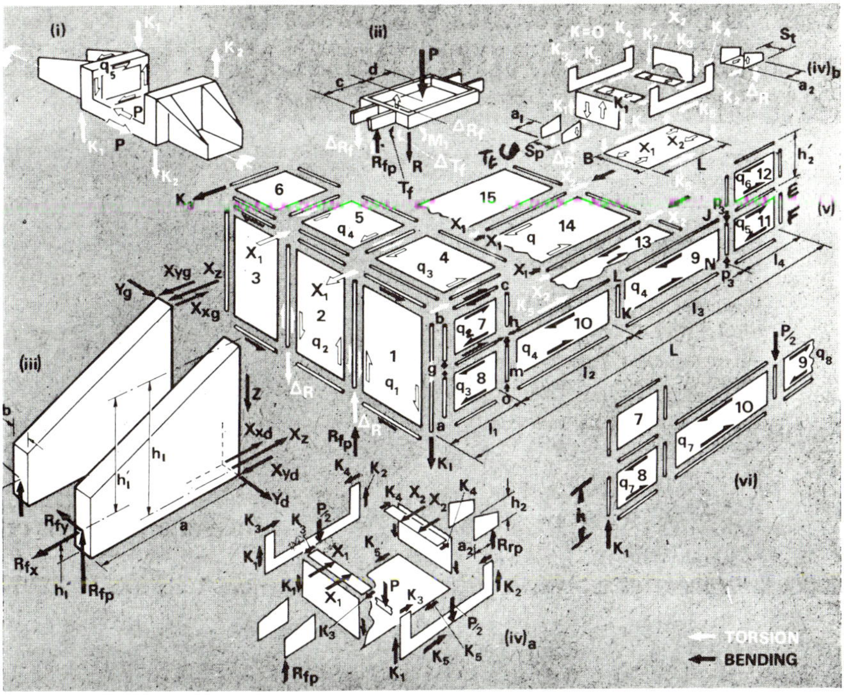

Fig. 4.11. Shear flow distribution in idealized punt structure

$q_5 = (q_6 h_2' - q_4 h)/h$ and the force balance on a–m–N–F gives $q_3 l_1 + q_4(l_2 + l_3) - q_5 l_4 - k_5 = 0$.

$P/2$, K_1, and K_2 forces: K_1 causes in panel 8, $q_7 = K_1/h$, part balanced by $P/2$ in L–K; then q_8, in panel 10, $= P/2h - q_7$. Summing the shear flows, for the front bulkhead panel in bars A–B, $q_1 = K_1/h_1$ which is transferred across panel 1 to balance R_{fp} and panel 2 is left unloaded. In the sidewall, groups of forces produce shear flows which are summed algebraically. Panels 8 and 11 with bars g–h and J–E have a concentration of internal forces and are critical to the structure.

TORSIONAL LOADING

Under this condition as in Fig. 4.11(iv)b there is little torsional resistance and forces on certain edges cannot arise and others have to be introduced into the floor. Wheel input loads R become edge forces on the bulkhead $= T_t/S_p$ and $K_1 = T_t S_p/B$.

X forces on the 'dash shelf' and floor are $X_1 = \Delta R a_1/h_1$

at the front, and at the rear $K_2 = \Delta R S_t/B$ and $X_2 = \Delta R a_2/h_2$.

These produce K_3 and K_4 on the sidewall: $K_3 = X_1 S_p/B$ and $K_4 = X_2 S_t/B$. K_5 arises from the sum of K_3 and K_4 and the floor accepts K_1.

Since K_1 in *a–b* causes $q_1 = K_1/h_1$ in panel 1, there is a balancing shear flow on the member carrying ΔR of $q_2 = (\Delta R - q_1 h_1)/h_1$. Similarly in panel 4, $q_3 = K_3/l_1$ and in panel 5, $q_4 = (X_1 - q_3 l_1)/l_1$; then in the floor $q_5 = K_5/L$ in panel 13 and $q_6 = (X_1 - X_2 + q_5 L)/L$ in panel 14.

The torsional weakness would be overcome by boxing in the front bulkhead to obtain torsional strength by differential bending of the sidewall. This creates a condition of static indeterminacy (see below) unless the structure is assumed symmetrical about the vehicle's transverse centre line.

CLOSED INTEGRAL CAR STRUCTURE

Structural surface idealizations for closed integral car structures require stiff jointed beam representation to allow for such surfaces as the side panel door pillars, windscreen, and backlight surrounds. Such stiff jointed frameworks are normally redundant structures, which implies the presence of more than one (statically determinate) load path between external load(s) and reactions. A redundant (or statically indeterminate structure) will require knowledge of the elastic properties of the members to establish the load distribution. By contrast, a determinate structure simply requires a knowledge of the geometry and kinematics of the structure to establish the load distribution.

Since real structures contain an infinite number of alternative load paths they are infinitely redundant so will always require idealization for analysis to a workable number of redundancies. A door framework (side frame) for a four door saloon shown at Fig. 4.12 has six degrees of redundancy visualized by cutting the two rings and counting the fixing reactions.

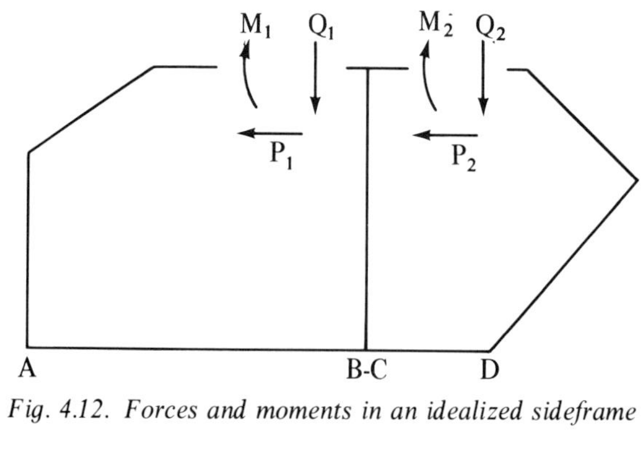

Fig. 4.12. Forces and moments in an idealized sideframe

The Unit Load method of analysis

Deflections of frameworks are conveniently analysed by the Unit Load method which can be used for both pin-jointed frames and continuum structures. For the second category, E. Goodwin in his D.A.T.A. booklet *The Unit Load Method of structural analysis* establishes the method by considering a small elastic element of an arbitrarily curved member under a generalized load system as in Fig. 4.13. This method is simpler to apply than the parallel 'unit displacement' method which is complex in that it involves the kinematics of the structure.

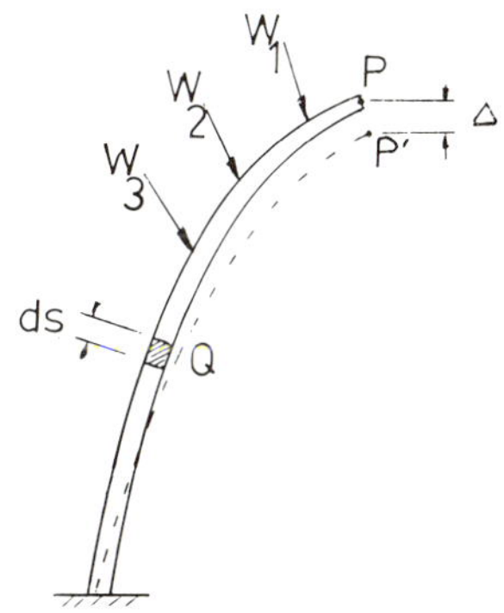

Fig. 4.13. Deflection of a continuum structure to be analysed by the unit load method

The structure is first considered infinitely rigid apart from the elastic element and the structure is unloaded save for a small vertical imaginary load w at P creating a bending moment at Q of M_0. On application of the real W loads, QP rotates through $\mathrm{d}\theta$ causing P to move vertically by Δ; the imaginary load w would thus do work w contributing $M_0\ \mathrm{d}\theta$ strain energy in the element. Therefore $w\delta = M_0\ \mathrm{d}\theta$.

The next stage is to allow for the elasticity of the whole structure so that

$$w\Delta = \int M_0 \,.\, \mathrm{d}\theta$$

so that

$$\Delta = \int \frac{M_0}{w} \,.\, \mathrm{d}\theta$$

Now M_0/w is the bending moment due to a load of unity $= m$, so

$$\Delta = \int m\ \mathrm{d}\theta \qquad (4.1)$$

And for R and R_0 local radius of curvature, and $\phi \to \phi_0 = \theta$ change of slope after and before bending, elastic modulus E and second moment of area I, then

$$\frac{M}{I} = \frac{E}{R} - \frac{E}{R_0} = E\frac{\mathrm{d}\phi}{\mathrm{d}s} - E\frac{\mathrm{d}\phi_0}{\mathrm{d}s} = E\frac{\mathrm{d}\theta}{\mathrm{d}s}$$

And $\mathrm{d}\theta = (M/EI)\,.\,\mathrm{d}s$ which on substituting in (4.1) gives

$$\Delta = \int \frac{Mm}{EI}\,.\,\mathrm{d}s$$

defining the structural deflection.

Redundant structures

Note that the unit load is dimensionless and must be made to act in the direction in which the structural deflection is required to be known. In using the expression for obtaining values of the redundant reaction in statically indeterminate structures the redundant reaction is regarded as the unknown applied force which produces zero deflection. The expression can be generalized for obtaining torsional, axial, and shear deflections (where they are significant) in

$$\int \frac{Mm}{EI}\,.\,\mathrm{d}s + \sum \frac{Pp}{AE}\,.\,\mathrm{d}s + \int \frac{Qq}{GJ}\,.\,\mathrm{d}s + \int \frac{Ff}{AG}\,.\,\mathrm{d}s$$

for member force due to external load P, due to unit load p; and member torque due to external load Q, due to unit load q; and shear force due to external load F, due to unit load f; elastic modulus E; shear modulus G; section second moment of area I; polar moment G; and cross-section area A.

As an example, to find the force W to produce 1·57 in (40 mm) vertically downward deflection at A in the determinate structure shown in Fig. 4.14

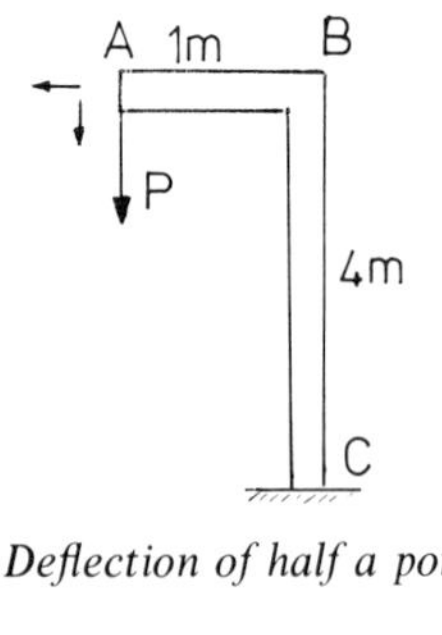

Fig. 4.14. Deflection of half a portal frame

given $E = 200\ \text{GN/m}^2$, $G = 80\ \text{GN/m}^2$, $I = 150 \times 10^8\ \text{m}^4$, and cross-section area $0{\cdot}004\ \text{m}^2$, allowing for bending, axial, and shear deformation

$$\delta_A = \int_0^l \frac{Pp \,.\, \mathrm{d}s}{AE} + \int_0^l \frac{Mm\, \mathrm{d}s}{EI} + \int_0^l \frac{Ff \,.\, \mathrm{d}s}{GJ}$$

$$= \int_0^4 \frac{-W \times -1\, \mathrm{d}s}{0{\cdot}004 \times 200 \times 10^9} + \int_0^1 \frac{W \times s \times s\, \mathrm{d}s}{200 \times 10^9 \times 150 \times 10^{-8}}$$

$$+ \int_0^4 \frac{1 \times W \times 1\, \mathrm{d}s}{200 \times 10^9 \times 150 \times 10^{-8}} + \int_0^1 \frac{W \times 1\, \mathrm{d}s}{0{\cdot}004 \times 80 \times 10^9}$$

$$= \frac{W}{10^4}\left[\frac{4}{80\,000} + \frac{13}{3 \times 2{\cdot}15} + \frac{4}{2{\cdot}15} + \frac{1}{32\,000}\right] = \frac{14{\cdot}443W}{10^6}$$

Member values:

	Pp	Mm	Ff
AB	00	$w \times s$	$w \times 1$
BC	$-w-1$	$1 \times w$	00

So for $\delta = 40$ mm,

$$w = \frac{40 \times 10^{-3} \times 10^6}{14{\cdot}443} = 2{\cdot}77\ \text{kN}.$$

For a redundant structure the Unit Load method can be used and where unknown internal forces are assumed for equilibrium and compatibility, conditions of zero displacement corresponding to the unknown forces are employed.

Thus for a portal frame under single central load P, Fig. 4.15, the unknown force and moment H and M are induced and by symmetry these are equal for each foot and the third pair of forces $V = P/2$. Only half the

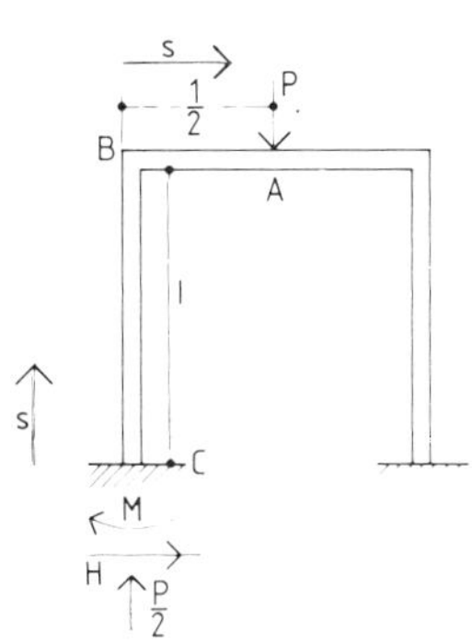

Fig. 4.15. Forces and moments in a symmetrical portal frame

frame need be analysed and the virtual deflections of C under the action of the reaction forces can be equated to zero to obtain M and H.

$$\delta = \int \frac{Mm}{EI} \, . \, ds$$

Member values:

	BC	AB
M	$M - Hs$	$M - Hl + \frac{p}{2} \, . \, s$
m_1	1	1
m_2	s	l

1. Rotation

$$\frac{1}{EI}\left\{\int_0^l (M - Hs) \times 1 \, ds + \int_0^{l/2} \left(M - Hl + \frac{p}{2} \, . \, s\right) . \, ds\right\} = 0$$

2. Horizontal displacement

$$\frac{1}{EI}\left\{\int_0^l (M - Hs) \times 1 \times s \, ds + \int_0^{l/2} \left(M - Hl + \frac{p}{2} \, . \, s\right) \times 1 \, ds\right\} = 0$$

1. Gives

$$Ml - \frac{Hl^2}{2} + \frac{Ml}{2} - \frac{Hl^2}{2} + \frac{Pl^2}{16} = 0$$

therefore 2 Hl − 3 M = Pl/8.

2. Gives

$$\frac{Ml^2}{2} - \frac{Hl^3}{3} + \frac{Ml^2}{2} - \frac{Hl^3}{2} + \frac{Pl^3}{16} = 0$$

therefore $5\,Hl - 6\,M = 3\,Pl/8$ which on solving simultaneously gives $H = P/8$ and $M = Pl/24$.

Another method of idealizing a sideframe, this time a pillarless saloon, is to consider A post, D post, and cantrail to be a portal frame as in Fig. 4.16 with joints B and C assumed pinned. The car body is in torsion with shear flow q along BC causing $2qh$ horizontal force in BC, bending AB and CD. Letting F equal the force on AB so $2qh - F$ acts on CD

$$\text{Deflection at B} = \frac{1}{EI}\int_0^h Fx^2 \, . \, dx = \frac{Fh^3}{3EI}$$

$$\text{Deflection at C} = \frac{1}{EI}\int_0^h x^2(2qh - F) \, . \, dx = \frac{2qh^4}{6EI} - \frac{Fh^3}{6EI}$$

Since these deflections are equal then $Fh^3/2EI = qh^4/3EI$ so $F = 2qh/3$. Thus bending moment at $A = 2qh^2/3$ and the relative stiffness of AB and

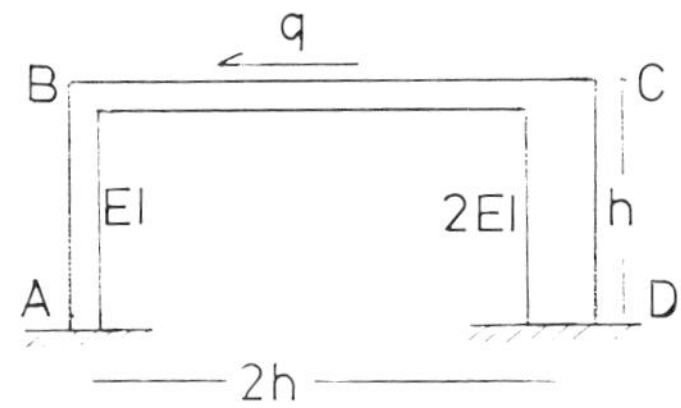

Fig. 4.16. Asymmetrical portal frame under shear flow q

CD are such that force at C is twice that at A, or $= 4qh/3$, giving moment at $D = 4qh^2/3$.

After finding deflections of various elements in a structure under a given load, these can be expressed as stiffnesses k or flexibilities f and combined in a way appropriate to their place in the structure. If the elements are in series the overall flexibility is the direct sum of the individual flexibilities so

$$f = \frac{\delta}{P} = f_1 + f_2 + \cdots \quad \text{and} \quad \frac{1}{k} = \frac{1}{k_1} + \frac{1}{k_2} + \cdots$$

If the structural elements are in parallel

$$k = k_1 + k_2 + \cdots \quad \text{and} \quad \frac{1}{f} = \frac{1}{f_1} + \frac{1}{f_2} + \cdots$$

BUS BODY IDEALIZATION FOR ANALYSIS

The bending deflections of bus sideframes, have been considered by Guy Tidbury ('The Structural design of bus bodies' (*Inst. Mech. Engrs/ASAE symposium* 1970)) who examines the effective second moment of area I of the body sideframe as used in the expression for deflection of a beam, under a central concentrated load of unit value, $\delta = L^3/48EI$. I is assumed to lie between a value computed for the whole frame section depth and that for the continuous panelling below the waist rail, I'_s. For average window pillar spacing Tidbury gives a value of $I_s = 1{\cdot}7\ I'_s$ taking into account the ability of the window pillars to transmit a certain degree of shear.

In considering sideframes with door openings, $1{\cdot}7\ I'_s$ is also used for the side panel beam either side of the door opening, with a section over the door opening subject to shear deflection only. Application of the unit load method to the two sections l_1 and l_3 in Fig. 4.17 joined by a rigid section over the door length l_2 gives

$$\delta_{\text{beam sections}} = \int_0^{e/2} \frac{x^2}{4EI_s} \, . \, dx + \int_0^{l_3} \frac{x^2}{4EI_s} \, . \, dx + \int_{l_2+l_3}^{e/2} \frac{x^2}{4EI_s} \, . \, dx$$

$$= \frac{1}{12EI_s} \left\{ \frac{e^3}{4} - (l_2 + l)_3)^3 + l_3^3 \right\}$$

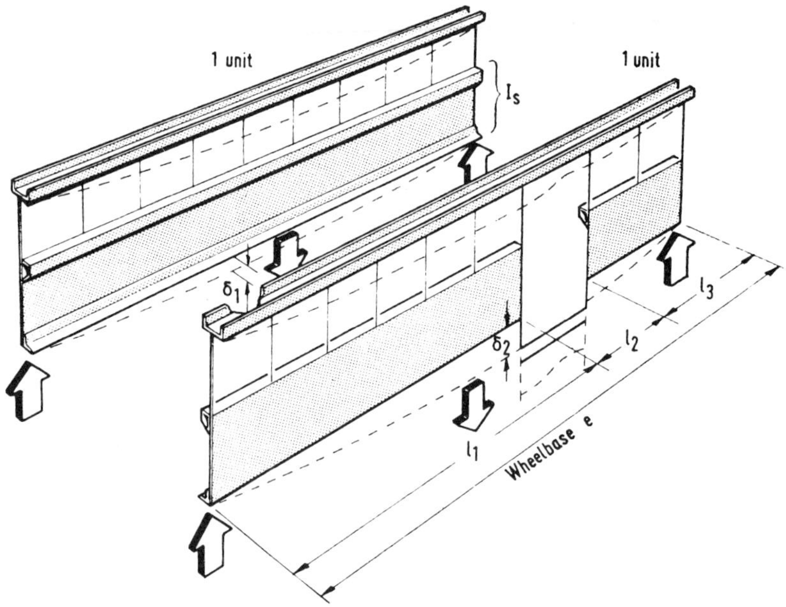

Fig. 4.17. Idealized bus body sideframes

Deflection due to shear over the door opening is considered as that of a fixed-ended beam with transverse end-constraint removed so $\delta = Fl/12EI$ where F is force tending to shear the beam equal to half the central load and $I = I_u + I_L$ (for cantrail and sole bar) so $\delta = (1{\cdot}5)^2 \times l_2^3/24E(I_u + I_L)$. A further factor of 1·5 on the bending moment allows for the beams belonging to a closed ring.

Figure 4.18 shows the overall bending of an asymmetrical bus body; two unit loads are applied to the centre of the vehicle, the overall flexibility will therefore be half the deflection given by these loads. Due to the shear stiffness of the structure joining the two sides, each sideframe will react a portion of a unit load according to its stiffness. If δ_1 is deflection of side with no door under force F_1 and δ_2 that of the doorway side under F_2 then shear deflection $\delta_3 = \delta_2 - \delta_1$. If the shear flexibility is f_3,

$$\delta_3 = (F_1 - F_2)f_3 = F_2 f_2 - F_1 f_1$$

or

$$F_1 = \frac{2(f_3 + f_2)}{2f_3 + f_2 + f_1} \quad \text{since} \quad F_2 = 2 - F_1 = \frac{2(f_3 + f_1)}{2f_3 + f_2 + f_1}$$

Centre deflection is

$$\delta = \frac{\delta_1 + \delta_2}{2} = \frac{F_1 f_1 + F_2 f_2}{2} = \frac{(f_3 + f_2)f_1 + (f_3 + f_1)f_2}{2f_3 + f_1 + f_2}$$

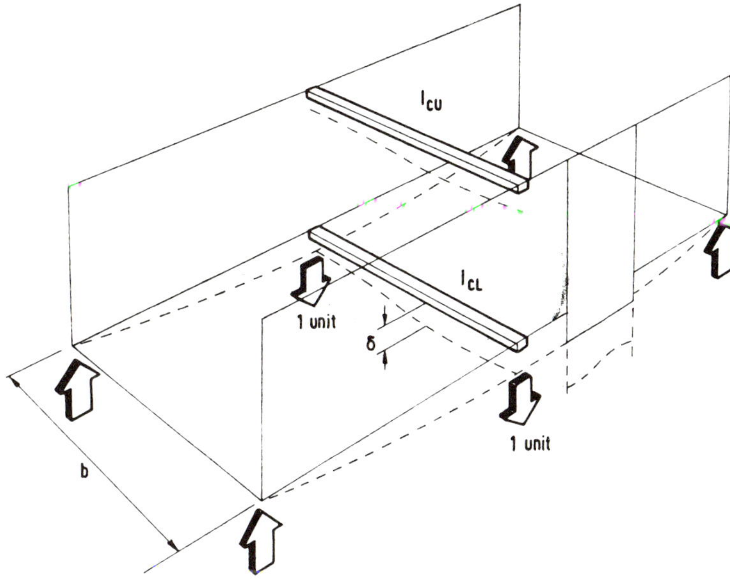

Fig. 4.18. Overall bending of an asymmetric bus body

and

$$\text{Overall flexibility} = \frac{\delta}{2}$$

$$= \frac{f_1 f_2 + f_3\left(\dfrac{f_1 + f_2}{2}\right)}{2f_3 + f_1 + f_2}$$

Bus body in torsion

K. Erz in his article 'Torsion in road vehicles caused by surfaces irregularities' (*A.T.Z.* no 4, 6, 11, 12, 1957) has developed a simple method for obtaining the torsional stiffness of bus bodies. He assumes that the side walls carry the main loads and that the structure below the waist rail is infinitely stiff compared with the beam elements surrounding the windows and doors. The structure is treated as a thin-walled tube with its axis across the vehicle, the transfer of shear from the cantrail to the waist rail putting the window pillars in bending; see Fig. 4.19.

The bending moment at any point along the rth window pillar will be $(2Q/3) \, . \, Q_{tr} \, . \, x_r$, treating the window pillars as fixed end beams and taking

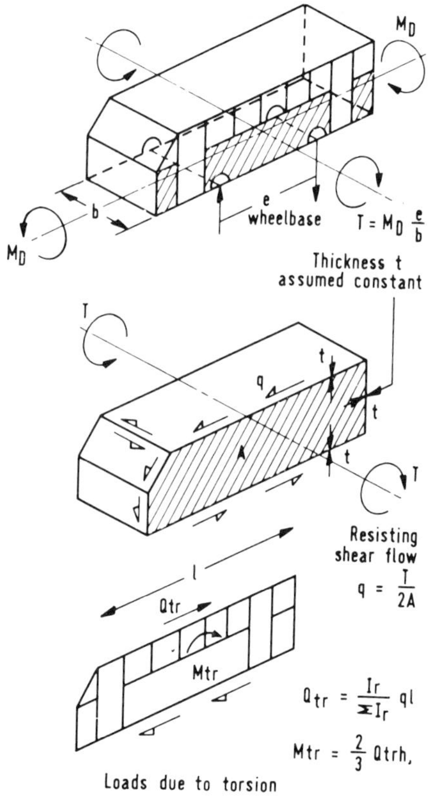

Fig. 4.19. Idealized bus body in torsion

Q_{tr} as the shear force acting on the rth pillar due to unit value of applied torque T to the body $T = M_D \,.\, e/b$

$$Q_{tr} = \frac{I_r \,.\, e \,.\, l}{\sum I_r \,.\, 2A \,.\, b}$$

since $T = 2qA$ and

$$ql = \frac{M_D \,.\, e \,.\, l}{2Ab}$$

Applying the unit load method, the angular deflection due to this pillar

$$\theta_{\text{pillar}} = \frac{1}{EI_r} \int_0^{h_1} \left(\frac{2}{3}\right)^2 \cdot \frac{I_r}{\sum I_r} \left(\frac{el}{2Ab}\right)^2 \,.\, x^2 \,.\, \mathrm{d}x = \frac{e^2 l^2 I_r \,.\, h_1^3}{27A^2 b^2 E(\sum I_r)^2}$$

Which on summing for all the pillars gives

$$\theta = f_t = \frac{e^2 l^2 h_1^3}{27A^2 b^2 E \sum I_r}$$

CHAPTER 5

THE VEHICLE AND ITS ENVIRONMENT: SAFETY AND FATIGUE

STRUCTURAL DESIGN FOR OCCUPANT PROTECTION

After weight saving the greatest influence on vehicle structural layout is that of the improved safety design for occupant protection. Besides designing in specific elements as high energy absorbers a much more detailed knowledge of the overall structural behaviour is required.

Normal technique in structural safety design is to ensure maximum strength and rigidity of the passenger compartment, structural deformation to absorb energy restricted to the front and rear ends, in the case of a family car. The experimental Sigma safety car in Fig. 5.1 is evidence of this approach while also using the tapered front end of the passenger compartment to deflect the power unit assembly downwards on impact.

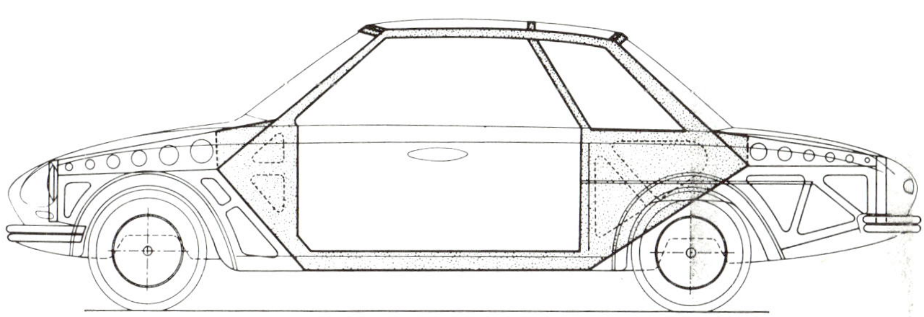

Fig. 5.1. Pininfarina Sigma safety car structural proposal

Crashworthiness criteria have been established and the following guidelines, from the United Nations, give those that should be adopted in design: WP 29/372, 373, 463, 477, 478, dealing in turn with front impact, fire risk, side, rear, and roof impact. Frontal impact requirement is that the passenger compartment should retain its shape after a 30 miles/h (13·4 m/s) solid barrier impact. In roof impact the windscreen pillar has to withstand a force equivalent to 60 per cent of the vehicle kerb weight.

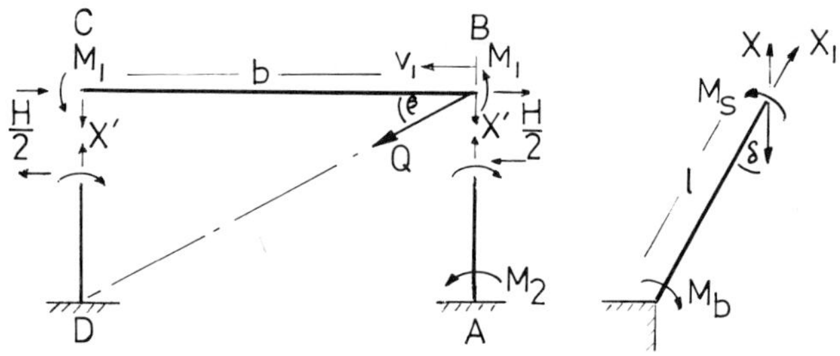

Fig. 5.2. Loads on windscreen frame on overturning

Forces in roll-over

W. Reidelback writing in *A.T.Z* for March 1964 ('Forces in overturned vehicle') has examined the roll-over case in relation to the windscreen frame comparing 'sloping' with 'dog-leg' A-pillars by an elastic analysis; for the structure in Fig. 5.2 under diagonal load Q he gives moment $M_1 = k \times Q \times l$ where

$$k = \frac{\cos \beta}{4 + [\{2(EI)_w \,.\, b\}/\{3(EI)_b \,.\, l\}]}$$

where EI_w is the stiffness of AB and EI_b is the stiffness of BC.

$$M_2 = Q.l.(\tfrac{1}{2} \cos \beta - k)$$

$M_5 = \mu M_6$ depends on the sideframe moment distribution

$$M_6 = l \,.\, \sin \delta \,(Q \,.\, \sin \beta - 2k \,.\, Q \,.\, l/b)/(1 + \mu)$$

Lateral displacement of top corner

$$v_1 = \tfrac{1}{2}Q \,.\, \cos \beta \frac{l^3}{3(EI)_w} - \frac{M_1 \,.\, l^2}{2(EI)_w}$$

M_6 and M_2 are acting on perpendicular axes so the resultant $M = M_2^2 + M_6^2$.

Head-on impact

The front end crush characteristics of vehicles affects the severity of impact for occupants and determines their movement, after impact, within the vehicle. So the way in which the vehicle front end structure is designed has a bearing on the degree of injury sustained by the occupants whether the vehicle is the 'bullet' or 'target' in the front-end impact situation.

The basic dynamics of impact have been studied by Grime and Jones in their paper 'Car collisions—the movement of cars and their occupants in accidents' (*Proc. Inst. Mech. Engrs* (AD) 1969–70 **184**, part 2A). They have put on record the fundamental equations describing the behaviour of idealized vehicles. In this study the vehicles (cars) are idealized in the sense of being assumed to have constant size of 13·75 ft (4·19 m) long × 5·25 ft (1·6 m) wide and weighing 2350 lb (1067 kg) including occupants. Studies of European and American car dimensions by the authors showed length to width ratios falling between 0·35 and 0·41 for a large sample, 0·38 being the most common.

Observations of front-end barrier impacts by the authors showed crush distances up to three feet (0·915 m) for 35–40 (15–18 m/s) impact speed with deceleration durations of about 0·1 s. In real impacts some 20 per cent of the mass (engine/transmission) may be brought to rest before the remainder of the car, an effect which is neglected. Also neglected is the influence of the tyre forces during impact and the effect of pitching and rolling motion. With $M_1\,V_1$ and $M_2\,V_2$ being masses and initial velocities of cars, before impact, which collide head-on in an assumed inelastic mode, the common velocity after impact

$$V_3 = (M_1 V_1 + M_2 V_2)/(M_1 + M_2)$$

As represented in Fig. 5.3(a) with co-ordinates x_1 and x_2, at the start of impact these are taken to be x_{10} and x_{20} and at the end of impact $x_{1\tau}$ and $x_{2\tau}$. The forces of the two cars are taken as $-X$ and $+X$ varying in time during impact as shown for a hypothetical impact pulse in Fig. 5.3(b).

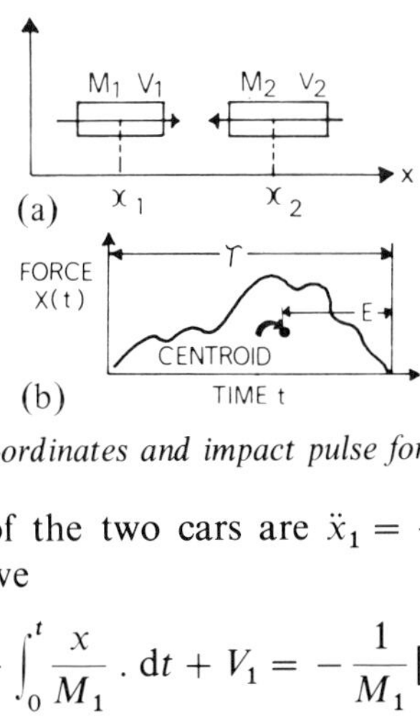

Fig. 5.3. Co-ordinates and impact pulse for colliding bodies

Equations of motion of the two cars are $\ddot{x}_1 = -X/M_1$ and $\ddot{x}_2 = X/M_2$ which on integrating give

$$\dot{x}_{1t} = -\int_0^t \frac{X}{M_1}\,.\,dt + V_1 = -\frac{1}{M_1}[A_x]_0^t + V_1$$

so that

$$x_{1t} = -\frac{1}{M_1}[A_x]_0^t\,.\,\bar{t} + V_1 t + x_{10}$$

and

$$x_{2t} = \frac{1}{M_2}[A_x]_0^t \,.\, \bar{t} + V_2 t + x_{20}$$

$\bar{t}$ is the difference between the time co-ordinate of the area under the Xt curve up to time t.

It can be shown that the distance moved by the centre of gravity of the cars

$$x_{1\tau} - x_{10} = V_1 \tau - \frac{1}{M_1}[A]_0^\tau \,.\, \bar{t} = V_1(\tau - \bar{t}) + \frac{M_1 V_1 + M_2 V_2}{M_1 + M_2} \,.\, \bar{t} \qquad (1)$$

$$x_{2\tau} - x_{20} = V_2 \tau + \frac{1}{M_2}[A]_0^\tau \,.\, \bar{t} = V_2(\tau - \bar{t}) + \frac{M_1 V_1 + M_2 V_2}{M_1 + M_2} \,.\, \bar{t} \qquad (2)$$

where $[A]_0^\tau$ is the area under the force/time curve which is the impulse given to each car.

The combined crushing of the two cars is equal to the difference between the distances moved by their centres of gravity which is $(V_1 - V_2)\times(\tau - \bar{t})$. The shape of the deceleration × time curve and its duration τ can determine $\tau - \bar{t}$. So for a given pair of cars the relative velocity at impact determines the characteristics of the impact. Figure 5.4 shows different kinds of pulse shapes for different front-end structure collapse modes with corresponding acceleration-distance/deceleration-distance ratios, giving $\tau - \bar{t}$.

The impulses given to the cars are equal and opposite and have the

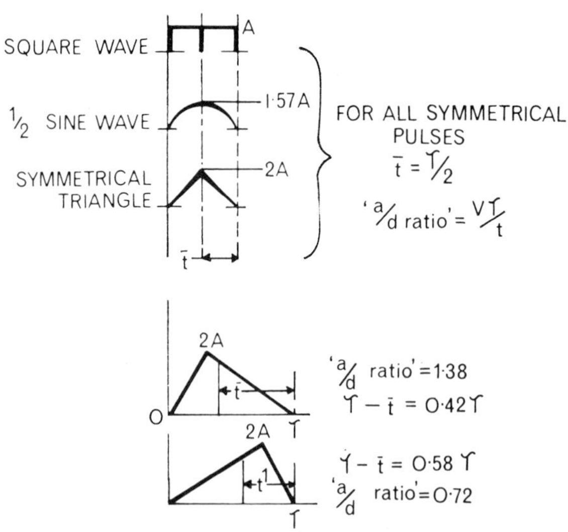

Fig. 5.4. Symmetrical and asymmetrical idealized impact pulses

value: $M_1 M_2(V_1 - V_2)/(M_1 + M_2)$ hence the deceleration/time curves of the passenger compartments are similar in shape but their magnitudes are inversely proportional to their respective masses. The areas under the deceleration/time curves, which are the changes in velocity of the cars, are

$$M_2(V_1 - V_2)/(M_1 + M_2) \quad \text{and} \quad M_1(V_1 - V_2)/(M_1 + M_2)$$

Energy loss during impact

$$= \tfrac{1}{2}M_1 V_1^2 + \tfrac{1}{2}M_2 V_2^2 - \tfrac{1}{2}(M_1 + M_2)V_3^2$$

$$= \frac{\tfrac{1}{2}M_1 M_2}{M_1 + M_2}(V_1 - V_2)^2$$

showing that the energy depends only on the masses of the cars and relative impact velocity so is unaffected by structure. However, with a crushable front end maximum energy absorption in a given distance is obtained so that the passenger compartment is protected from intrusion at as high as possible impact velocity.

PLASTIC COLLAPSE

The plastic overstressing and collapse of beams was discussed in Chapter 3, in defining the form factor for a beam. When the full plastic moment M_0 is applied to the beam and the yield stress has been reached over the entire cross section a plastic hinge develops at the point of highest moment and the beam undergoes large permanent deformation.

For a simply supported beam length L subjected to a central load, collapse occurs when the load P is $4M_0/L$ compared with $8M_0/L$ if the beam is fixed at its ends. So this would be the point of collapse if an impulsive force of this magnitude was applied. For higher values of P each half of the beam rotates as a rigid body in addition to the rigid body acceleration of the whole beam. When $P = 45{\cdot}6M_0/L$, plastic hinges also form at points $0{\cdot}202L$ either side of the centre and for higher P move inwards towards the centre.

Impact energy in collision is equal to load × deformation; aim in design is to use as much (permanent) deformation as is allowable without encroachment on the driver's zone of clearance. Rails and pillars can be arranged so that plastic hinges occur at their points of intersection. Collapse mode after the hinges have formed is based on the structure acting like a hinged mechanism.

The collapse mode can be modified by gussetting of joints to alter the effective lengths of the arms of the mechanism; presence of shear panels will also affect the mode of collapse. Different structural criteria apply in design for crashworthiness because plastic collapse requirements will be affected (in the formation of hinges) by the neutral axis to extreme fibre dimension rather than the square of its dimension as in elastic design. There is a tendency to specify smaller but more heavy gauged rails with stiffer joints.

For a conventional car structure with open engine well at the front end, the structure surrounding the engine compartment comprises radiator and front panel with the two wings and valances, and forward extensions of the underbody side rails: this section can be idealized as a portal frame rooted at the bulkhead. Figure 5.5 shows such a frame under front and side loads V and H with three alternative collapse modes which will depend upon the geometry and relative stiffness of the structural elements, as well as the load pattern. Plastic moments are M_{ps} for the stanchions AB and DE and M_{pb} for the beam BD; at corners B and D the smaller plastic moment is M'_p.

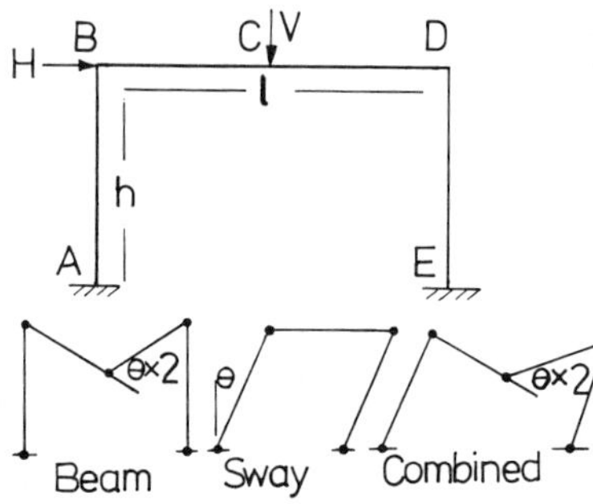

Fig. 5.5. Plastic collapse modes for portal frame

In the beam collapse case, work done by the central load is $V \, . \, \frac{1}{2}l\theta$ and equal to the energy dissipated in the plastic hinges $M_{pb}\,2\theta + 2M'_p\,\theta$ so $\frac{1}{4}Vl = M_{pb} + M'_p \ldots$ (a); similarly for sway collapse $\frac{1}{2}Hh = M'_p \ldots$ (b), and for combined collapse $\frac{1}{4}Vl + \frac{1}{2}Hh = M_{pb} + M'_p \ldots$ (c). These apply if the portal frame is assumed to have pinned feet; if fixed feet are assumed (b) becomes $\frac{1}{2}Hh = M_{ps} + M'_p \ldots$ (b)$_f$ and (c) becomes $\frac{1}{4}Vl + \frac{1}{2}Hh = M_{ps} + M'_p + M_{pb} \ldots$ (c)$_f$.

A typical car front-end portal 4·92 ft (1·5 m) wide and 4·92 ft (1·5 m) high (long) on a vehicle subject to 2225 lbf (10 kN) central load simulating a post impact, with a lateral component of 565 lbf (2·5 kN)

$$\tfrac{1}{2}Hh = \tfrac{1}{2} \times 2{\cdot}5 \times 1{\cdot}5 = 1390 \text{ lbf/ft } (1{\cdot}875 \text{ kNm})$$

$$\tfrac{1}{2}Vl = \tfrac{1}{4} \times 10 \times 1{\cdot}5 = 2765 \text{ lbf/ft } (3.75 \text{ kNm})$$

If (a) and (c)$_f$ are satisfied $M_{ps} = 1{\cdot}875$ and for $M'_p = M_{ps}$, $M_{pb} = 1{\cdot}875$. If (b)$_f$ and (c)$_f$ are satisfied $M_{pb} = 3{\cdot}75$, $M_{ps} = 0{\cdot}9375$.

Choosing the first solution, required section modulus $Z = M_p/S\sigma_y$ where

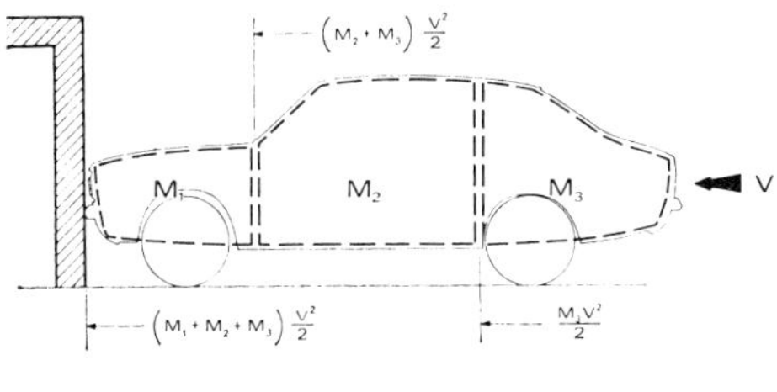

Fig. 5.6. Three mass idealization of impacting vehicle

S is the shape factor and σ_y is the yield stress for stanchion and beam.

Analogue methods

The criteria for design of the front-end structure have been considered by Finch, in a paper read at the UK Transport and Road Research Laboratory conference in January 1972, who states that the objective is to strike a compromise between excessive stiffness resulting in high passenger compartment decelerations and excessive flexibility giving rise to steering column penetration and other mechanical intrusion into the occupant space. Finch portrays the vehicle as a system of separate masses, Fig. 5.6, the energy to be absorbed by each division diminishing from front to rear.

Power unit location well forward to allow space in front of the dash bulkhead and to relieve the adjacent structure of absorbing the kinetic energy of the engine, as well as that of the whole vehicle, was a prime consideration. The next requirement was tuning of impact load carrying panels, such as wings and valances to produce a progressive collapse sequence front to rear. Static simulation of this pattern of structural collapse is obtained in the crush rig shown in Fig. 5.7, in which links and spreader bars are used in conjunction with hydraulic rams to effect realistic load distribution to the major mass centres.

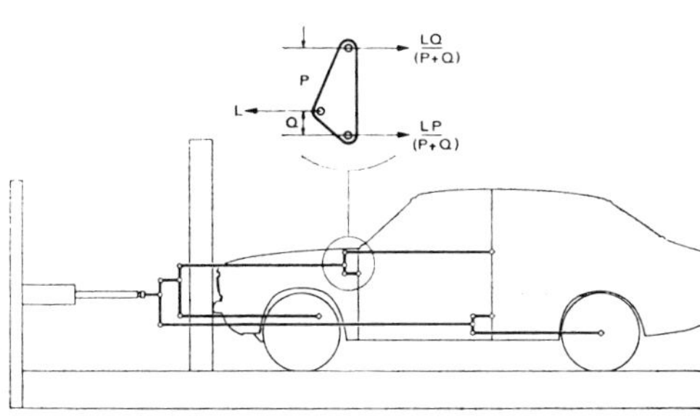

Fig. 5.7. Static rig to obtain crush characteristics

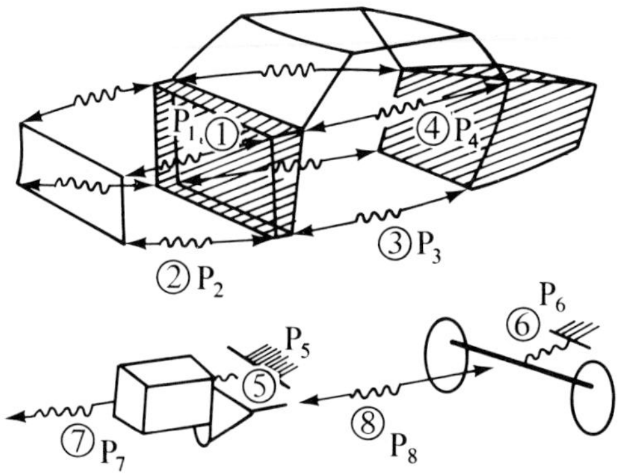

Fig. 5.8. Mathematical model for impacting body shell

Miura and Kawamura writing in *SAE* paper 680484 have shown how they have constructed spring/mass analogues to the car structural characteristics for studying impact behaviour in simulation, and determining an optimum balance of strength between the structural members. The analogue model is shown in Fig. 5.8 with the load deformation characteristics of the structural member simulated as shown in Fig. 5.9 and a diagrammatic arrangement of the model in Fig. 5.10. Equations of motion for the system were solved step-by-step from the initial conditions with the aid of a digital computer and results were obtained within 5 per cent of experimental ones, to illustrate the reliability of the method.

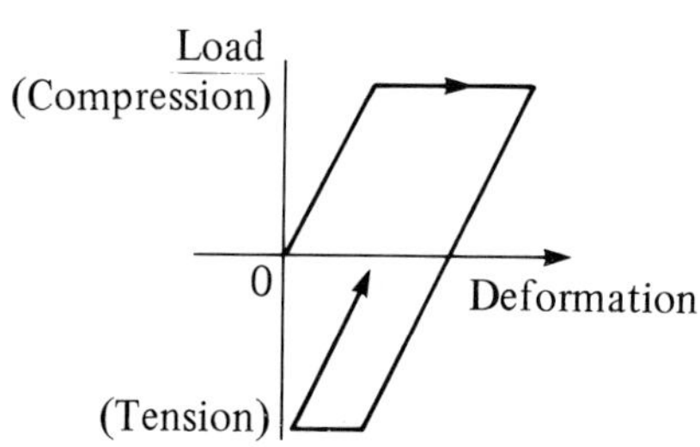

Fig. 5.9. Deformation diagram for modelled body member

Similar techniques have been used by Emmerson who used the histories of crush-test characteristics obtained by Finch (in papers read at the UK Transport and Road Research Laboratory conference in January 1972) to predict performance of new designs by correcting static crush results for inertia and strain rate effects in the computer programme; experimental and computer simulation results are shown as Fig. 5.11.

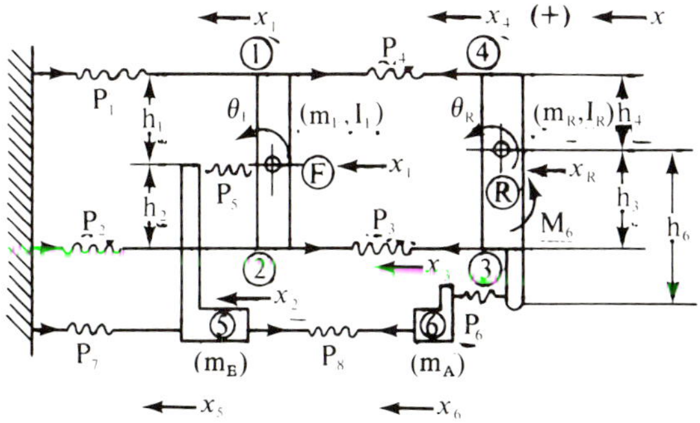

Fig. 5.10. Schematic mathematical model for impacting body shell

Workers in the USA have shown how deflectors designed to divert the engine beneath the passenger compartment served to extend the crushable length of the front end during impact. Modifications to a production saloon car structure are shown at Fig. 5.12. An unmodified car of the same type crashed against a pole at 60 mile/h (26·8 m/s) and recorded peak deceleration of $85g$; the modified car recorded a $55g$ peak. In addition to the deflector the modifications comprised collapsible struts connecting the reinforced bumper (giving increased strength in the early stages of crush) to the passenger compartment on two levels. The upper strut gave way at 40 000 lbf (177 kN) and the lower at 60 000 lb (256 kN).

Controlled collapse designs

Barracos and Rhodes writing in SAE paper 690004 have been able to predict the collapse rate of front-end structures which are specially designed to be energy absorbent. In the arrangement shown in Fig. 5.13 curved beam theory predicts controlled collapse of the S-shaped forward longerons (made from 50 000 lbf/in^2 (344 MN/m^2) UTS steel of 0·132 in (3·35 mm) thickness) in the region of the curve, subject to moment arm $l_m = 4{\cdot}7$ in (11·95 cm). Radius of the neutral axis $R = A/\int_a \mathrm{d}A/r$ for cross-section area A and r is the radius to desired stressed (σ) fibre. To evaluate

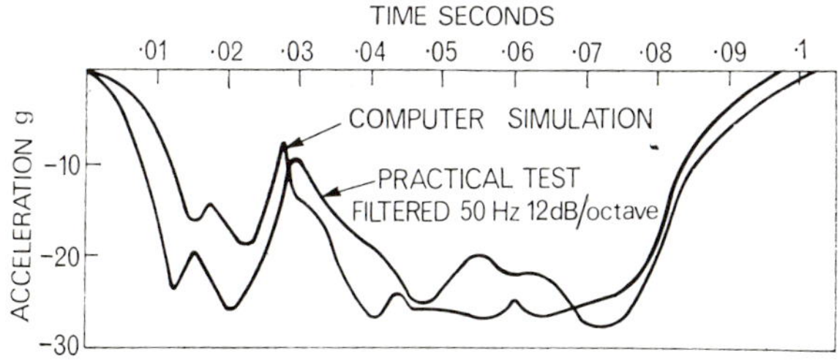

Fig. 5.11. Simulated and measured impact pulses

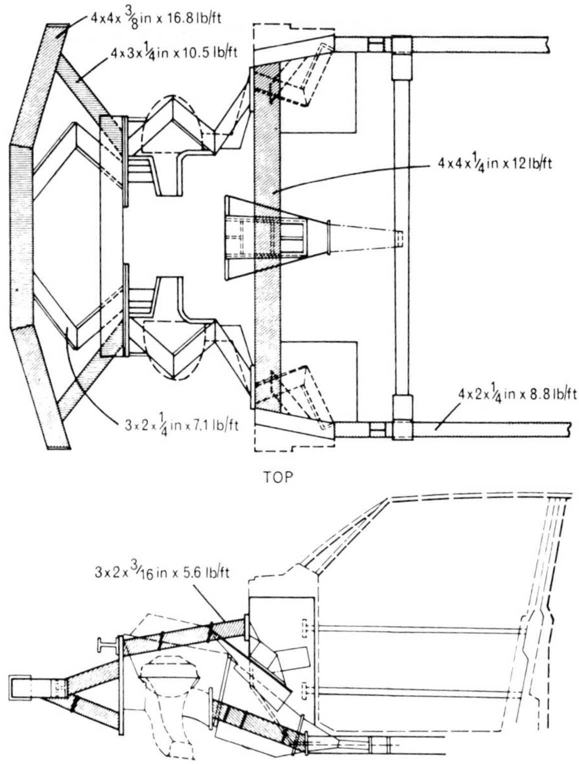

Fig. 5.12. Controlled-collapse girders for front-end structure

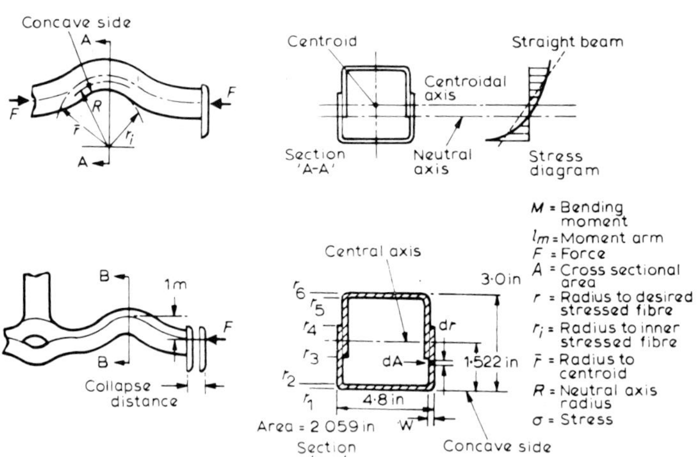

Fig. 5.13. Curved beam controlled collapse longeron

the denominator the cross section is divided into rectangular sections, width W, so that

$$R = \frac{A}{[(W_1 l_m r_2/r_1) + (W_2 l_m r_4/r_2) + (W_3 l_m r_5/r_3) + (W_4 l_m r_6/r_5)]}$$
$$= 5{\cdot}64 \text{ in } (14{\cdot}3 \text{ cm})$$

and so $r = 5{\cdot}92$ in (150 mm) and collapse force per side rail for yield $= \sigma \,.\, r \,.\, A \,.\, (r - R)/(R - r) \,.\, l_m = 21\,700$ lbf (96·5 kN).

Practically straight members subjected to end load tend to bend because of unavoidable eccentricity of the applied load; see Fig. 5.14. As the axial load increases the lateral displacement grows at an increasing rate as indicated by the curve marked 'elastic'. As the displacement increases so does the bending moment at the centre and a plastic hinge will form at B given by

$$\left(\frac{P}{P_0}\right)^2 + \frac{4v_m}{d}\left(\frac{P}{P_0}\right) = 1$$

where $M_0/P_0 = d/2$ for a rectangular cross section which when plotted is the line marked 'plastic' in Fig. 5.14. If the member has a slender section it will become unstable at a compressive load producing stresses below the

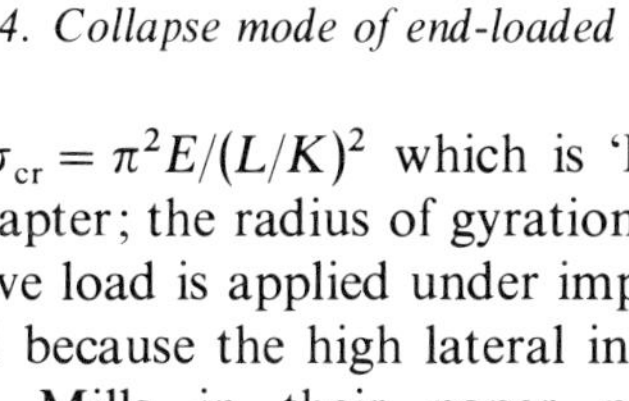

Fig. 5.14. Collapse mode of end-loaded member

yield stress given by $\sigma_{cr} = \pi^2 E/(L/K)^2$ which is 'Euler buckling' as considered in an earlier chapter; the radius of gyration $K = \sqrt{(I/A)}$.

When the compressive load is applied under impact conditions, the collapse load is increased because the high lateral inertia forces restrain the member. Barley and Mills in their paper presented at the 1970 ASAE/Inst. Mech. Engrs Summer conference on 'Body Engineering' have shown that the time for plastic hinges to form was reduced for higher velocity impact and that deformation before hinges formed was only a fraction of the total deformation. Energy absorption in this plastic stage was found to be independent of impact velocity, so that static collapse characteristics were applicable. Recent studies have shown that the collapse of thin-walled tubes under bending has more relevance to real cases of roll-over and side impact.

Crumpling of thin-walled sections under axial impact will occur only in the front part of the member because of longitudinal inertia effects. The expression for static crumpling load has been found to apply under impact conditions: $P = 6\sigma_y t\sqrt{(Dt)}$ for a circular section tube diameter D and thickness t (σ_y = yield stress). Crush length x may be found by equating kinetic energy of the mass M to the work done Px so that

$$x = Mv^2/(12\sigma_y t^{1\cdot 5} D^{0\cdot 5})$$

The Invertube

Means of extending the energy absorption characteristics of circular tubes have been described by Searle and Brabin in *M.I.R.A. Bul.* No. 2, 1970. Energy absorption can take place in causing a tube to turn back on itself until it is partly inside out; such a set-up, involving specially shaped mandrels or dies, is known as Invertube and can provide uniform deceleration in a square form pulse. In idealized form it is shown as Fig. 5.15.

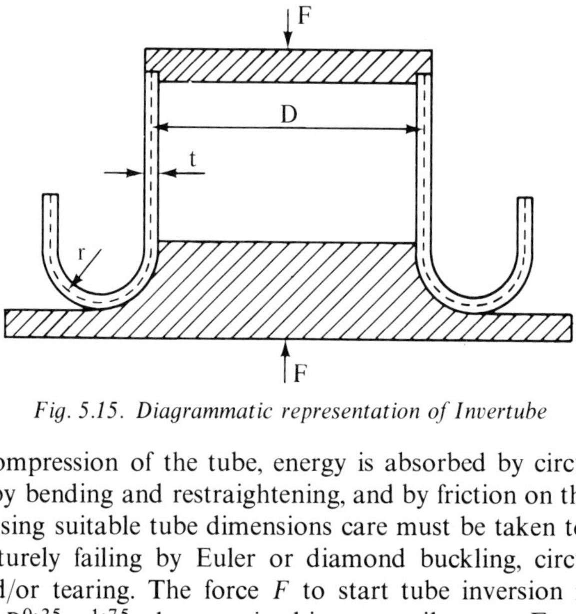

Fig. 5.15. Diagrammatic representation of Invertube

During compression of the tube, energy is absorbed by circumferential expansion, by bending and restraightening, and by friction on the inverting die. In choosing suitable tube dimensions care must be taken to avoid the tube prematurely failing by Euler or diamond buckling, circumferential bending and/or tearing. The force F to start tube inversion is given by $F = 25\,\sigma_{ult} \cdot D^{0\cdot 25} \cdot t^{1\cdot 75}$ where σ_{ult} is ultimate tensile stress. For any material where σ_{ult} is not known it is convenient to measure Vickers Pyramid hardness number and divide it by 4·6. Thickness to diameter ratio must be chosen to avoid premature failure by other modes.

The collapse behaviour of rectangular-section tubes has been studied in the prediction of impact behaviour of bus bodies. Investigation of scale models under impact has been carried out by W. T. Lowe and colleagues writing in *J. Automotive Engng* January 1972; these authors used dimensional analysis to calculate realistic scaling and establish an expression for crush resistance. The following relationships need to be equal for model and full size structure

$$\frac{x}{L}, \frac{\sigma_y}{\rho V^2}, \frac{b}{L}, \frac{ab}{V^2}, \frac{t}{b}$$

for crush length x, vehicle length L, yield stress σ_y, density ρ, impact velocity V, vehicle width b, acceleration a, and material thickness t.

Collapse of a simple rectangular-section tube is given by crumpling resistance in

$$P \,.\, x = \frac{V^2}{2g}(1 + c)\rho AL$$

for cross-section area A and ratio of non-structural to structural mass c and vehicle (weight W) retardation $a = P \,.\, g/W$.

The crumpling load P is obtained in

$$\frac{P}{P_0} = \left(\frac{6A}{A_0} + 0{\cdot}04\right)\left(1 + \frac{V}{50}\right)$$

where $P_0 = \sigma_y \,.\, A$; A_0 is the area enclosed by A so that A/A_0 is the 'solidity ratio'.

Models were made to check these expressions and then to simulate effects of windows, underbody structures, and other items so that factors of variance could be established.

FATIGUE AND VIBRATION

So far we have considered structures under the effects of either steady or impact loading but not of repetitive cyclic loading. Elements of a structure have been designed to operate within the elastic limit, or to meet a certain rigidity requirement, unless designed for plastic collapse under impact load. As well as plastic collapse another form of failure which should be considered is that of fatigue.

Fatigue failure can occur at stresses much smaller than the elastic limit, under cyclic load conditions. The failure normally commences at a discontinuity or a surface imperfection such as a small crack which is propagated under cyclic loading until it spreads across the section and leads to rupture. Even with ductile materials failure occurs without generally revealing plastic deformation.

The stress intensity may be completely reversed cyclically or fluctuate around a mean constant value; Fig. 5.16 shows the terminology for describing stress level. The fatigue life of the structure (or element) is the

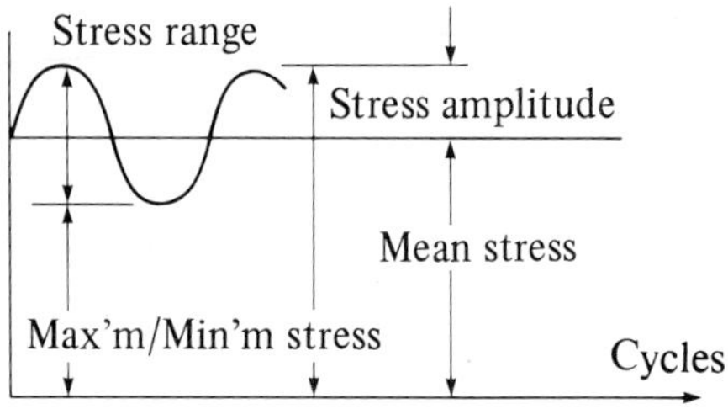

Fig. 5.16. Fatigue nomenclature

number of cycles of stress sustained prior to failure. A plot of stress against number of cycles to failure is called an S–N diagram and helps to describe the fatigue resistant properties of different materials based on endurance tests which have been carried out. They often define an endurance range of limiting stress, say, on a 10 million cycle basis.

Plotting stress against cycles to failure on a logarithmic scale reveals a straight line graph to describe an exponential relationship such as $S = CN^{\alpha}$ where C and α are constants depending, respectively, on the material and the type of endurance test. The graph in Fig. 5.17 shows that there is a change of slope to zero at a given value for ferrous metals describing an absolute limiting range of stress for an indefinitely large number of cycles. For non ferrous metals no such limit exists and it is common practice to define a 'fatigue limit' of 5×10^8 (for aluminium alloys).

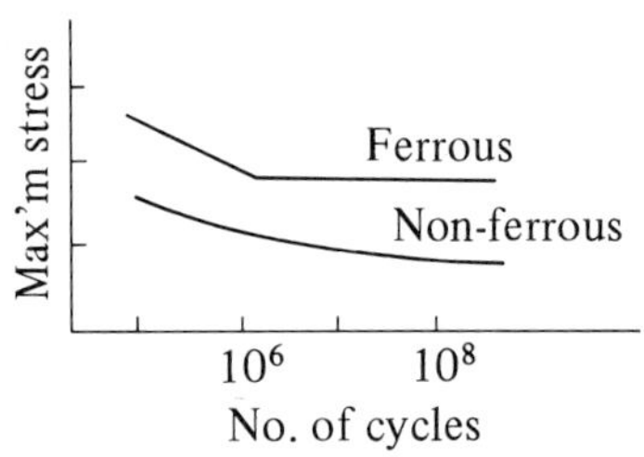

Fig. 5.17. SN curves for metals

With motor body steels it would therefore be possible to ensure that the fatigue limit was never exceeded and therefore build a structure with infinite life. In practice the structure would be too heavy so the design for a finite life has to be considered. In practice designs are proved by accelerated fatigue testing of critical structural elements.

Tests on box sections

S–N curves for typical mild steel box sections, of a similar size cross section to that of a sill or underbody side member, have been obtained by the British Welding Research Association (Papers FE 23/5/56, 23/7/58,

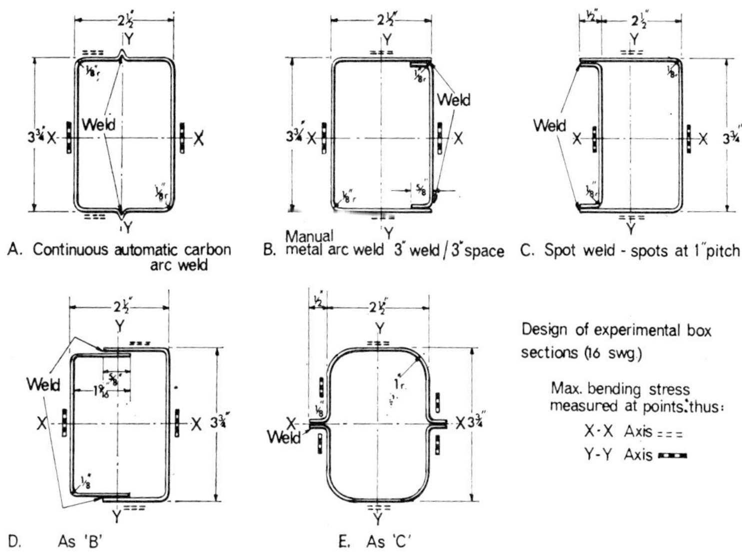

Fig. 5.18. Typical sections for vehicle rail box members

23/6/58). Five different fabricated sections, Fig. 5.18, each $2\frac{1}{2} \times 3\frac{1}{4}$ in ($6{\cdot}35 \times 8{\cdot}27$ cm) were made from 16 swg En 24 mild steel and subject to bending vibrations (with 'free-free' support conditions), and then torsional vibrations. The steel used had an ultimate tensile strength of 19·34 ton/in^2 (300 MN/m^2) and 47·28 per cent elongation.

Effective second moments of area I_{xx}, I_{yy} were obtained from static tests and are given in Table 5.1. Bending vibrations about XX and YY axis were excited in the range 3500–5000 c/min (53–83 Hz) and centre span stresses were recorded with strain gauges. Applied bending moment was recorded in SN curves as in Fig. 5.19 (a) and (b) for XX and YY axes. Section E proved superior probably because spot welds at the neutral axis were sufficiently strong under alternating shear. Deduced reduction factors for the other sections as a function of section E are shown in Table 5.2.

Polar moments of inertia, for use in the torsional tests, were again found by static torsional tests. For the dynamic tests, vibrations were excited by

TABLE 5.1

Section moments of inertia, in^4

Section	I_{XX}	I_{YY}	J
A	1·55	0·79	1·57
B	1·67	0·81	1·31
C	1·98	1·09	1·43
D	1·75	0·75	1·24
E	1·07	0·94	1·30

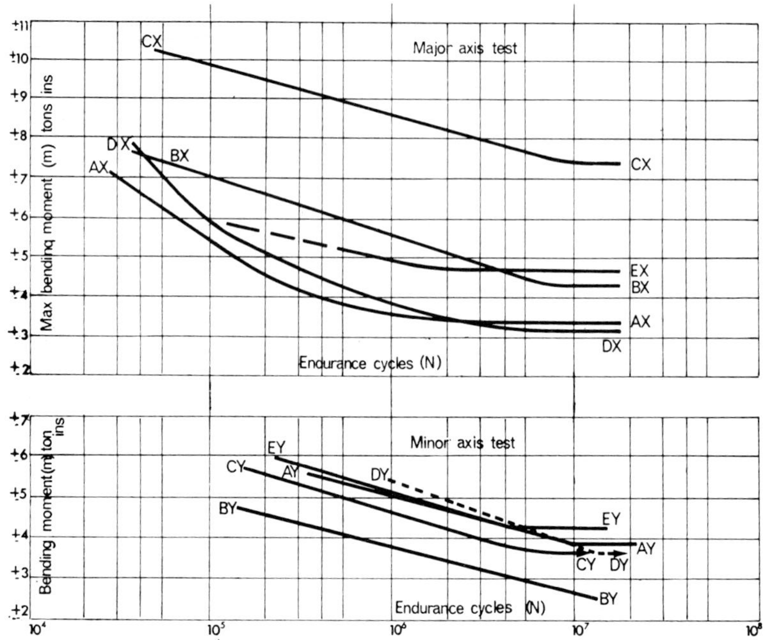

Fig. 5.19. SN curves for mild steel box beams

an oscillator at one end of the beam specimens while the other ends were fixed, and the beams vibrated at their torsional resonant frequency given by

$$\omega = \frac{60}{2\pi}\sqrt{\left(\frac{G \cdot J \cdot 12g}{I_0 \cdot L}\right)}$$

For shear modulus $G = 11{\cdot}7 \times 10^6$ lbf/in^2 (80·7 GN/m^2), length $L = 36$ in (0·916 m), oscillator inertia $I_0 = 3620$ lbf in^2 (10600 kg cm^2), tor-

TABLE 5.2

Strength reduction factors in bending

Specimens	*Fatigue strength at 10^7 cycles ($\pm$ tonf/in^2)*	*Strength reduction factor*
EX	8	1·0
CX	7	1·1
BX	4·5	1·8
AX	4	2·0
DX	3	2·7
EY	5·5	1·4
BY	4	2·0
CY	3·5	2·3

sion constant J for 'perfect' section = 1·82 in^4 (75·7 cm^4) and g = 32·2 ft/s^2 (981 cm/s^2), ω = 2400 c/min (40 Hz).

This time section A, with its continuous weld providing a fully closed section, gave the best results and was used as a standard for comparison in Table 5.3. Tests on 'actual' sill sections used in design studies could be carried out in a similar manner at the conceptual stage in design to be used in fatigue studies of idealized structures.

TABLE 5.3

Strength reduction factors in torsion

Specimens	*Applied torque corresponding to 10^7 cycles (tonf/in)*	*Strength reduction factor*
A	1·8	1
B	0·75	2·38
C	1·7	1·05
D	0·75	2·38
E	1·3	1·3

Structural vibration

Road input vibrations to the vehicle structure are generally of a random nature with variable frequency and amplitude content which have to be evaluated for transformation into accelerated laboratory test programmes. Sometimes road-induced and engine-induced vibrations excite different parts of the structure to vibrate in sympathy so an appreciation of structural vibration is useful.

As well as affecting fatigue life, structural vibration can affect both vehicle ride and handling. Generally it is possible to design enough torsional and bending stiffness into a vehicle structure so that first natural frequencies fall in a range of 20–30 Hz and there is little danger of these being excited by the suspension frequencies. Otherwise of course relative movement of the suspension anchorages would severely affect handling stability and control.

In terms of ride comfort, however, body structural vibrations excited by high frequency road inputs can cause difficulties. Some manufacturers put body and suspension running gear in different vibrational environments by means of suspension subframes and perimeter chassis frames, soft mounted to the body to filter out ride 'harshness'.

The various modes of vibration of a car body have been illustrated by Jha and Priede, writing in *Automotive Design Engng* November 1972, as in Fig. 5.20. These authors found that the maximum level of 'noise' inside a car lies in the sub audible frequency range around 10 Hz and is caused by

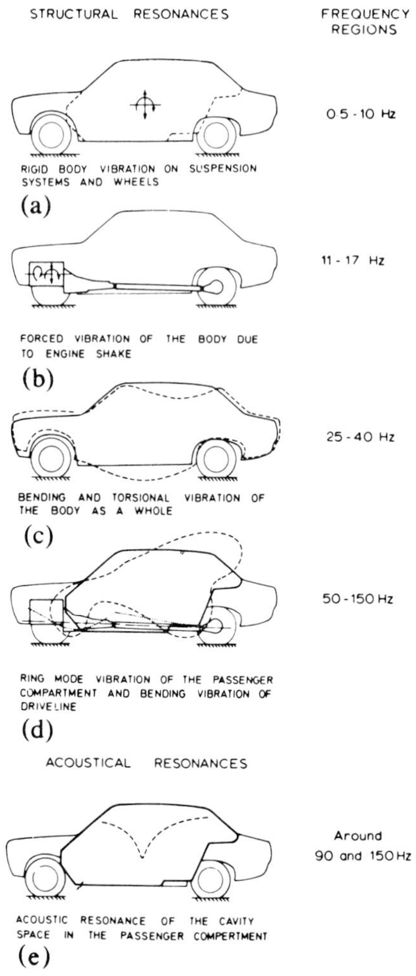

Fig. 5.20. Vibrational modes for saloon car body shell

modes (a) and (b) in the figure. In the audio range mode (c) and particularly mode (d) were important with cavity resonance (e) being of only secondary importance.

A rough approximation to the resonant bending frequency of a vehicle body, particularly of the open punt shell type, can be obtained by finding the second moment of area of the main side members I, and estimating an assumed equal mass per unit length then substituting in the following expression for natural frequency

$$f = 1{\cdot}57\sqrt{(EI/mL)}\ \text{Hz}$$

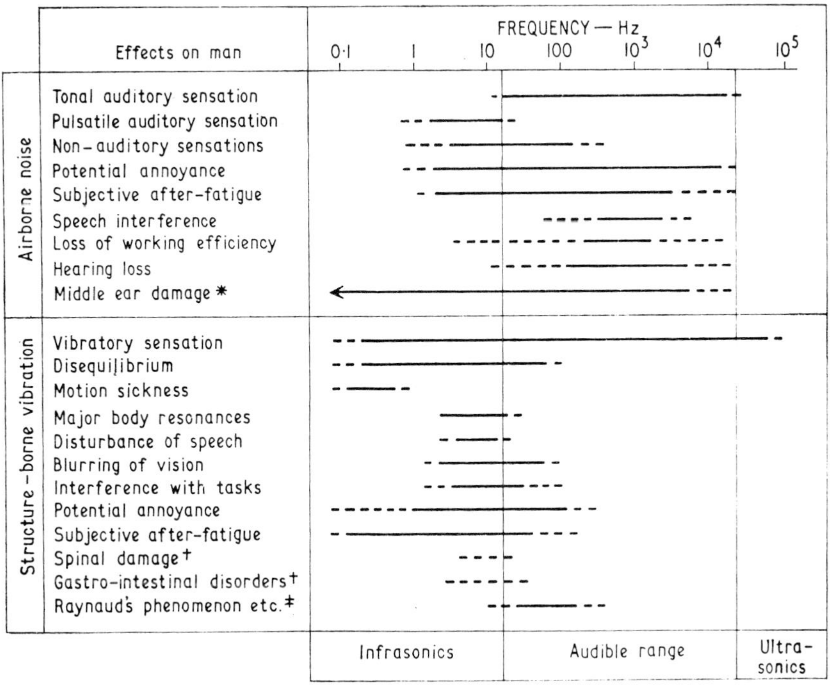

Solid lines show known frequency ranges; hatched lines show uncertain ranges of occurrence.

* Pressure injury (barotrauma) at extreme intensities.

† Chronic disorders associated clinically with repeated occupational exposure to whole-body vibration.

‡ Chronic diseases affecting the blood vessels, bones and joints of the hand and forearm associated with cumulative occupational exposure to vibration from hand-held power tools.

Fig. 5.21. Response of human behaviour to vibration

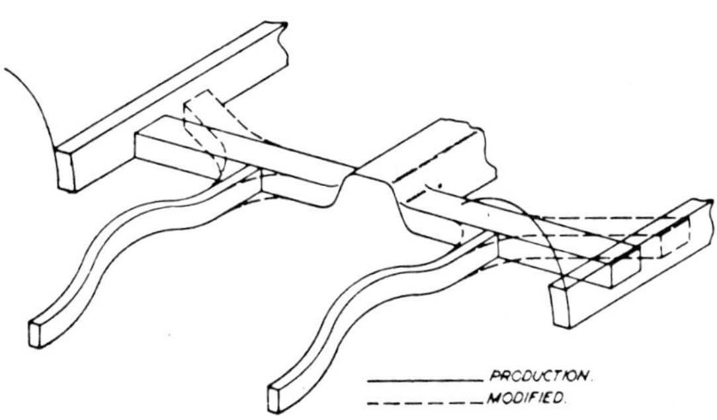

Fig. 5.22. Modified underbody structure to obtain high frequency vibration isolation

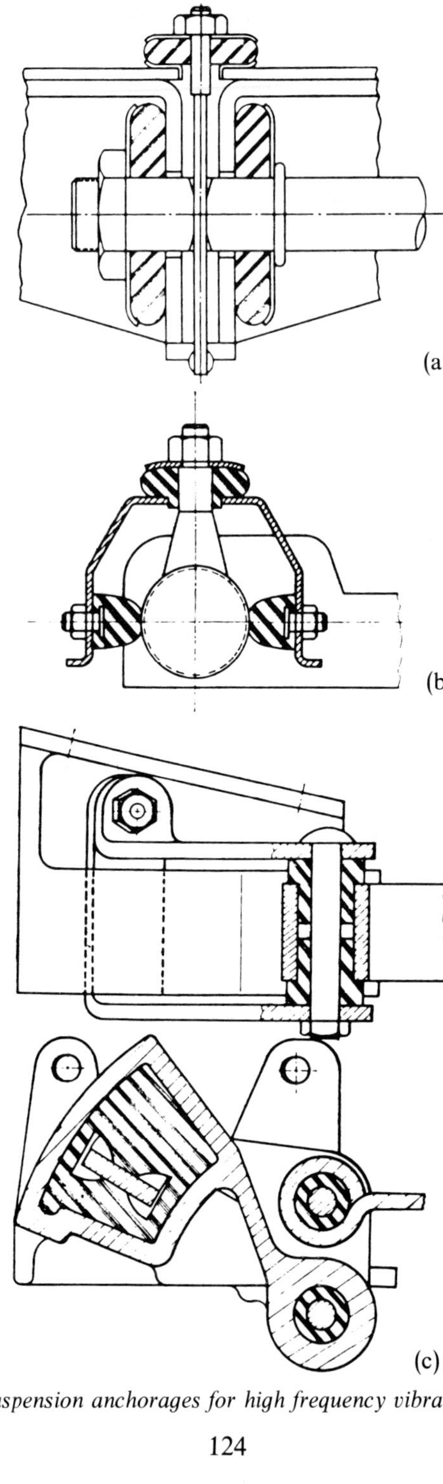
(a)
(b)
(c)

Fig. 5.23. Suspension anchorages for high frequency vibration isolation

where L is the wheelbase and E the body material elastic modulus and m is mass per unit length.

Response and compliance

The effects of vibrations on man have been categorized by Guignard in *Proc. Inst. mech. Engrs* 1967–8, **182**, Part 1 who shows the response of different body functions in terms of exciting frequency in Fig. 5.21. Factors other than frequency, such as intensity of stimulation and duration of exposure to it, are also important. In broad terms Guignard says that the human tolerance of whole body vibration is lowest in the frequency band from 4 to 8 Hz.

Vibration frequencies in the 0–5 Hz band are largely the province of the suspension designer lying as they do in the 'ride' frequency range. From 5–25 Hz, shake, and from 25–200 Hz, harshness, vibrations fall half in the province of the structural engineer and half with the suspension man.

Boden at the 1962 Summer ASAE conference on Vehicle Ride considered these vibrations and concluded that their isolation is perhaps more a job for the development engineer than the designer. A typical source of excitation is the joints between cast slab type concrete roads. Boden points out that tuning engine mounts to the unsprung mass natural frequency helps isolation in some vehicles. A more fundamental approach may be to tackle local 'vibrating elements' within the whole body structure. These might possibly reveal themselves as discontinuities in the plots of bending and torsional structural deflection along the length of the body, as referred to in Chapter 1.

For example, changes to the junction of rear side member, cross member, and outer sill on a saloon car shown in Fig. 5.22 reduced vibration. Other ways of reducing shake and harshness centre around attention to the suspension anchorages, introducing carefully controlled 'compliance'. The method of control needs careful experimentation as the *ad hoc* introduction of flexible rubber bushes could affect the handling motions of the vehicle. A method of introducing steeply rising rate characteristics for a rubber sandwich ball joint at the end of the drag strut of a McPherson-type independent front suspension is shown in Fig. 5.23(a). A laterally loaded tie wire design stiffens up the endwise compliance of the strut after initial deflection of the rubber 'doughnuts'. A design for a lower wishbone anchorage pivot is shown in Fig. 5.23(b) as another means of obtaining controlled compliance and in Fig. 5.23(c) is shown a rear spring front bush shackle design.

CHAPTER 6

STRUCTURE DESIGN FOR PRODUCTION

IN LARGE-SCALE vehicle manufacture the overriding constraint on body design is cost. Although this is a book on structural design rather than body manufacture it would be foolish to ignore the cost constraint since it applies at every stage in design. Thus when there are alternative structural solutions to a design problem the one that is least likely to be expensive in material, tooling, and assembly is clearly the most likely one to be adopted.

Value analysis techniques should be considered whenever possible by structural designers since they can lead to a fresh approach to a problem often resulting in a more efficient design, structurally as well as costwise. The technique is to take a clearly defined function such as: stabilizing a boot lid; reacting rear-suspension loads; anchoring a seat belt; then, without prejudice, comparing the total cost of a function with the competition.

GRAVIMETRIC ANALYSIS

D. G. Roberts of Ford, writing in *Automotive Design Engineering* November 1966, made a useful description of the cost versus function argument in defining 'design bulldozing' as a term applied to an engineer who despite all the implications or comparisons shown him, still refuses to accept that his design contains unnecessary details or costs. Conversely, some organizations permit a considerable amount of 'cost bulldozing' to the detriment of product function and a careful balance has to be found.

The first step is to establish a possible part-by-part breakdown of a product such as that in Fig. 6.1, the breakdown being in a form that allows a cost estimator to quickly determine piece price and tool cost. First review is usually a comparison with similar products for either a previous or competitive model. Value analysis is next applied and part-by-part, function-by-function costs compared. Thereafter a decision can be made as to whether less parts or simpler ones could perform the same function or whether certain product features can be dispensed with.

Component cost build-up can be defined as the sum of the following factors: total net cost of material, accounting for scrap reclamation, labour cost of operation relative to the production of the part, and fixed overhead being that section of the overhead that exists whether or not the part is in production. Variable overhead is the part of the overhead directly related

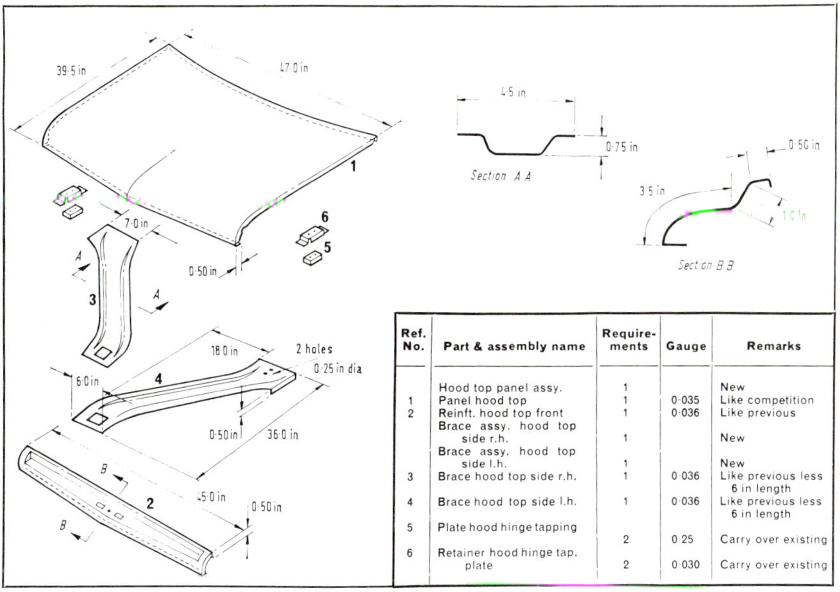

Ref. No.	Part & assembly name	Require-ments	Gauge	Remarks
	Hood top panel assy.	1		New
1	Panel hood top	1	0·035	Like competition
2	Reinft. hood top front	1	0·036	Like previous
	Brace assy. hood top side r.h.	1		New
	Brace assy. hood top side l.h.	1		New
3	Brace hood top side r.h.	1	0·036	Like previous less 6 in length
4	Brace hood top side l.h.	1	0·036	Like previous less 6 in length
5	Plate hood hinge tapping	2	0·25	Carry over existing
6	Retainer hood hinge tap. plate	2	0·030	Carry over existing

Fig. 6.1. Component make-up for hood top assembly

to making the part, being the cost that stops when production stops; it includes such items as machine running costs and material handling cost related to the part. Figure 6.2 shows the contributory components of a cost for the manufacture of a complete body shell, showing in particular the high material element. Besides having low first cost and reliability a part must also be capable of easy manufacture. It should be remembered that a skilled experimental mechanic might spend half an hour on an operation which when in full production might have a two minute cycle time in the hands of a relatively unskilled operative.

DESIGN FOR PRESSWORKING

The basic operation of blanking out metal shapes for subsequent press-forming can be made more economical in several ways. Avoid the temptation of radiused corners or unduly sharp ($< 45°$) corners; both can cause

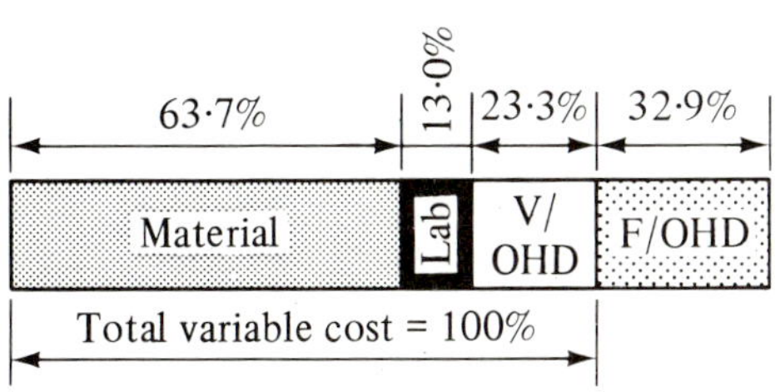

Fig. 6.2. Cost build-up for complete car body shell

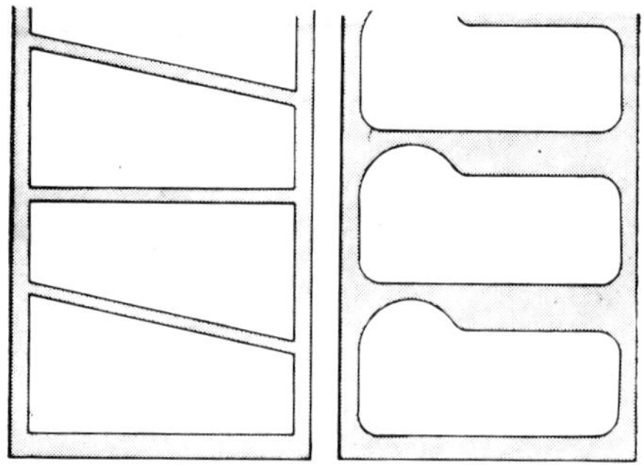

Fig. 6.3. Blank outline for efficient strip usage

premature punch replacement. The perfect blending of radii with straight lines makes an attractive drawing but Fig. 6.3 shows how it can increase scrap metal on the strip feed to the blanking press. This is because in strip used for progression blanking, the sharp cornered parts nestle more closely together and there is less overall scrap on the strip which otherwise contains more loosely arranged rounded-corner components.

Simple bending as shown in Fig. 6.4 can cause some problems, the minimum bend radius, for example, can range from 0·5t for soft material up to about 7t for full temper steels (where t is the metal thickness). These radii can go up to 100 per cent again if the bending is with the metal grain. In calculating the developed length in $l = b + c + (t/3 + r)\ 1{\cdot}5708$, note that this can vary with ductility and grain direction. Also the result would be correct for the 'open vee' tool (left) but not for the 'side-drawing' tool (right). And to keep down the bending tool cost in both cases allow a generous tolerance on the 90 degree angle, say, +2 degrees.

In designing parts for the drawing press bear in mind that the metal has

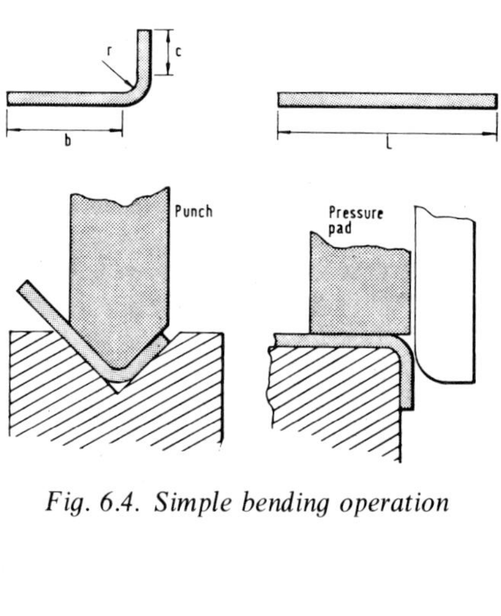

Fig. 6.4. Simple bending operation

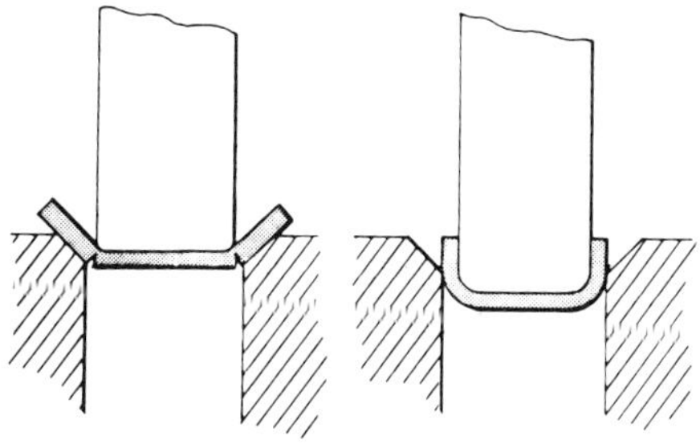

Fig. 6.5. Simple drawing operation

to flow, so radii at the bottom of the part should be as large as possible. If they are too small, see the result in Fig. 6.5; but with greater expenditure a sequence of drawing operations can be made to produce the fine radius if really required. Also about 2 degrees of draft (taper of the 'vertical' walls) is helpful on most work, to avoid the occurrence of wrinkling.

In drawing, remember that the most economical production is achieved when the depth of draw does not exceed $1\frac{1}{2}$ times the punch diameter. Otherwise more operations with successive tooling are required perhaps even with intermediate annealing.

When designing a pierced component, distance between holes and from holes to the edge of the component should be at least twice metal thickness, otherwise the wall thickness of the die cannot withstand the piercing pressure.

In deep drawing the change in shape which the blank undergoes should be considered. For making a cup from a circular disc blank, Fig. 6.6; R_1 must reduce to R_2, L_1 to L_2, and displacement of the shaded material sets

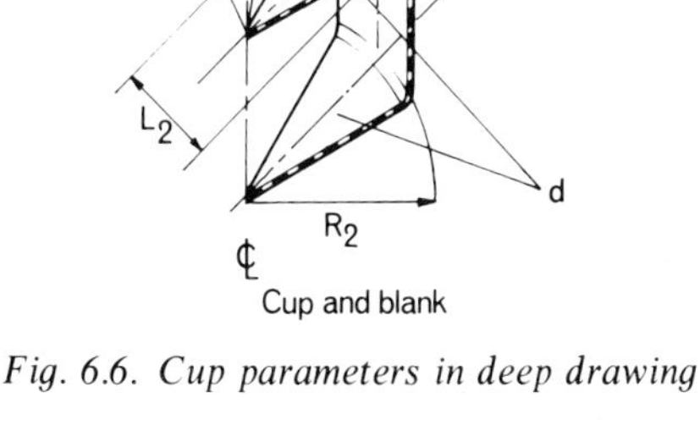

Fig. 6.6. Cup parameters in deep drawing

up hoop stress in the annular region outside the die aperture. The resulting strains influence the amount the blank diameter can be reduced, the effective norm being 40–45 per cent resulting in a cup height 60 per cent of its diameter.

A 'square' component can be drawn to a depth at least equal to its width providing the ratio of depth to corner radius is no more than five. For general asymmetric pressings where metal has to flow in several directions complicated stress patterns arise and a sharp change over the die edge can cause excessive strain on the blank and lead to die failure. Hence die aperture edge radii must be within 4 to 10 times metal thickness.

At the 1968 Institute of Sheet Metal Engineering conference A. H. Cole provided the following valuable hints for designing pressed parts; see Fig. 6.7.

(a) Flanging: Raising of a flange where the blank contour passes acutely across the bend line should be avoided; otherwise, as the metal diminishes in width towards the bend line, it fails to meet the die wall and an unsightly bulge results.

(b) Swaging: When beads, stiffening ribs, or depressions are required in a panel their root radius should be as generous as possible consistent with the strengthening effect required. Allow ample clearance around swages to allow possible easing of the radii; clearance may be necessary to alleviate cracking during proving of the tools.

(c) Wrinkling avoidance in shells: if a clean finish panel is required, work must be carried out at all points or the remaining, unworked, metal will remain soft and easily distort. For a straight drawn shell, an adequately strong bottom is only possible by stiffening in the form of beads and/or adopting a radiused die line at the closed end.

(d) Shells with long straight sides: Incorporate flanges where possible in long shells. This avoids distortion of the panel by releasing stresses during the trimming operation. It is worthwhile, too, if possible to design wall curvature, in plan, to counteract inherent wall curvature introduced by the die radius.

(e) Redesign of pedestal bearing for forming in sheet metal: At (*a*) a stiffening dish avoids a welding operation; at (*b*) auto turned bobbins are retained by spot-welded U-brackets.

(f) Do flanges have to be circular? If not, one can capitalize on continuous strip tooling by specifying flats to represent self-width strip. This lessens material usage, also allows dropping away the scrap at the rear of the press.

(g) Do flange widths always matter? This top-hat section trim valence is more expensive where parallel flanges are specified since different curves are developed at front and rear edges. By varying flange width it is possible to incorporate this blanking as a cropping operation in the forming tool.

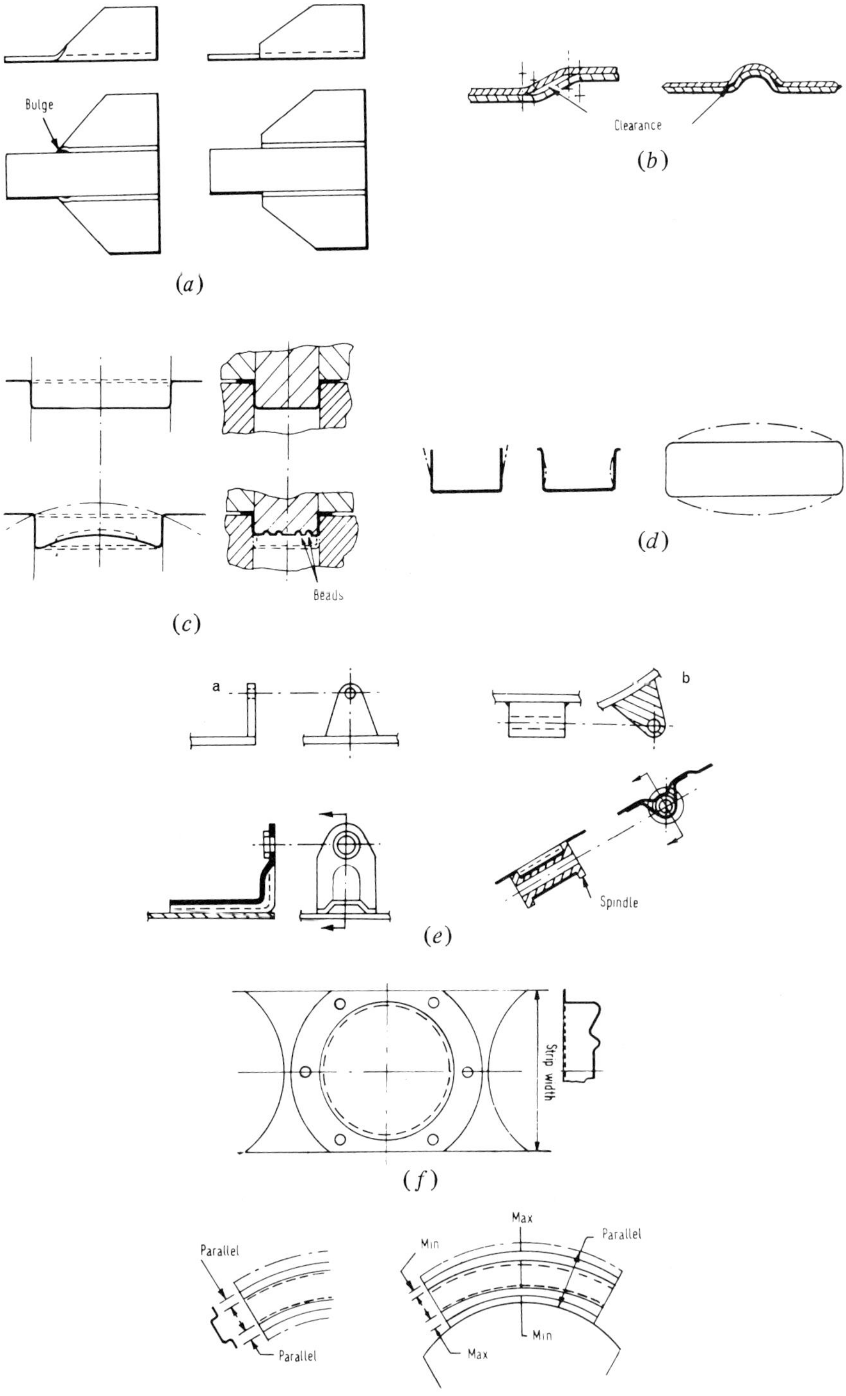

Fig. 6.7. Do's and Dont's in designing for presswork

DESIGN FOR SPOT WELDING

Production spot welding on complete body build or sub-assembly line is rarely up to laboratory or experimental shop standards so that generous factors of safety have to be allowed. Accessible joints make for better welding so designers should ensure that flanges to be joined are not excessively shielded by surrounding sheet metal work.

Welding dissimilar metal thicknesses can also cause difficulties and a 3 : 1 maximum differential for mild steel must be observed. Some preferred joint arrangements for particular duty are shown on the right hand side on each instance in Fig. 6.8. (a) is a baffle to avoid water entry, (b) is a 'slip' joint to cater for assembly variations, and (c) a self-locating flange joint. (d) is a joint that demands 'lineability' and should therefore be avoided where possible on visible surface applications, whereas (e) demonstrates ways of avoiding the need for spot welding more than two metal thicknesses.

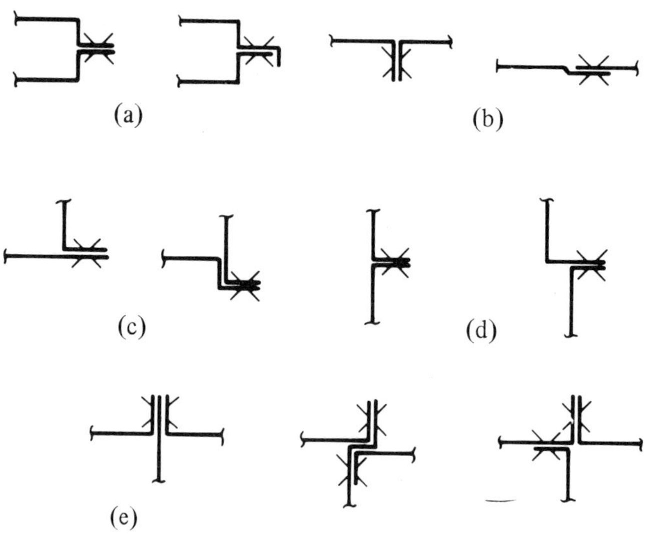

Fig. 6.8. Do's and Dont's in designing for spot welding

Joints must also be arranged to simplify weathersealing so that in, for example, the roof/cantrail part section of a saloon car, shown in Fig. 6.9, the drain channel is formed out of the roof panel and not, as is often the case, from the extension panel (cantrail to roof). Reinforcement of the heel board is another example of keeping water out of a closed section, by mounting the cross member *above* the floor pan.

Special attention must be paid to structural quality in the following areas where spot weld integrity and tolerance stacking on panel dimensions

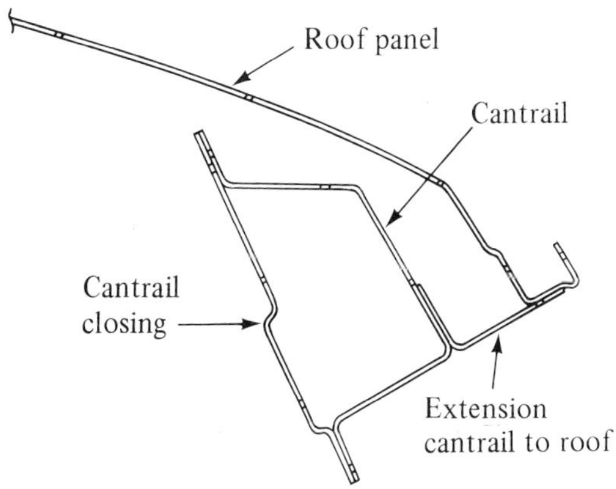

Fig. 6.9. Drip rail design to avoid leaking joint

will be vital; these are the unitary underframe, front valance–dash side–A post connection, rear wheelarch–sill–D post connection, and the B–C post.

SOME PRODUCTION PROBLEMS

Giles in his book *Vehicle body construction and design* (Iliffe, 1970) has considered the design of the B–C post, shown sectioned with adjacent door part section in Fig. 6.10. The requirement for a slender section above the waist rail requires the 'drop' of the door glass by the lock and hingeing areas to be established before the final width of the post. Sometimes it is possible to taper the glass drop from top to bottom in order to build up the effect of a slender upper part of the post. Depth of the post is first established by the swinging of the hinge and the rear door, and width is affected by the opening swing of the front door and the relationship with the lock and striker mechanism, and relative positions of the weatherstrip sealing surfaces. The depth of draw required could demand a multi-stage pressing and the problem of reacting the hinge loads may involve the addition of a stiffening liner. Figure 6.11 shows how the cumulative effect of adverse tolerances can have a bearing on design.

Sometimes designs are severely compromised by available production equipment particularly in respect of commercial vehicles for manufacture in underdeveloped countries. In SAE paper 975 B, 11-15, 1, 1965, Mazurek and co-workers describe the construction of the Chrysler XLV truck cab shown in Fig. 6.12 which was fabricated from sheet steel stock mainly with the use of a shearing machine, bending rolls, band saw, and pressbrake. The use of piano hinges simplifies the door post and hood design, and the

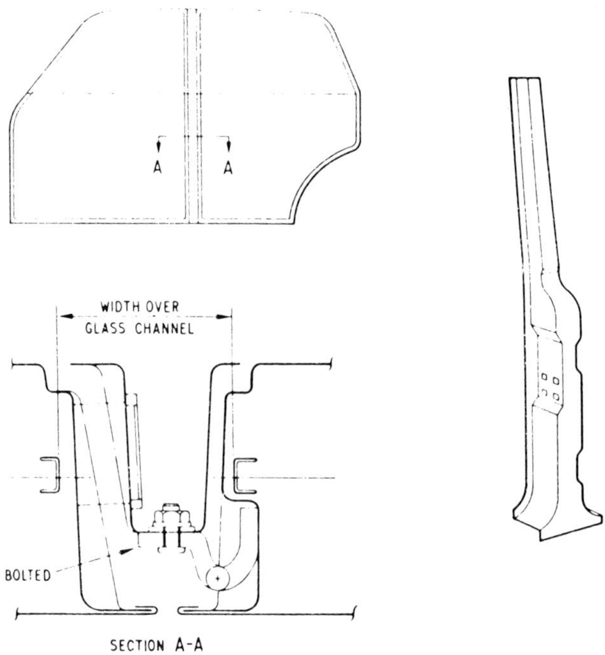

Fig. 6.10. B–C post section, hinge and glass detail

design of straight-walled sections considerably eased the forming process. Four metal gauges only were used of 0·042, 0·060, 0·090, and 0·120 in and there was minimum need for projections and generations of curved surfaces. Door flanges and sealing are unique and the section through the lower windscreen support shows the sturdy transverse box section employed. Because of the 'squareness' of the design, stress concentrations at

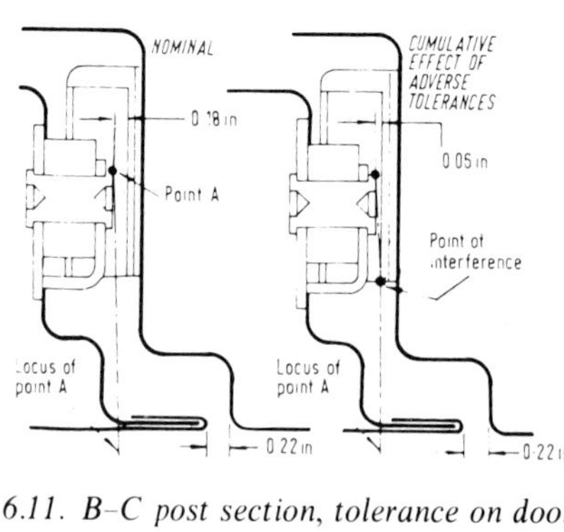

Fig. 6.11. B–C post section, tolerance on door latch

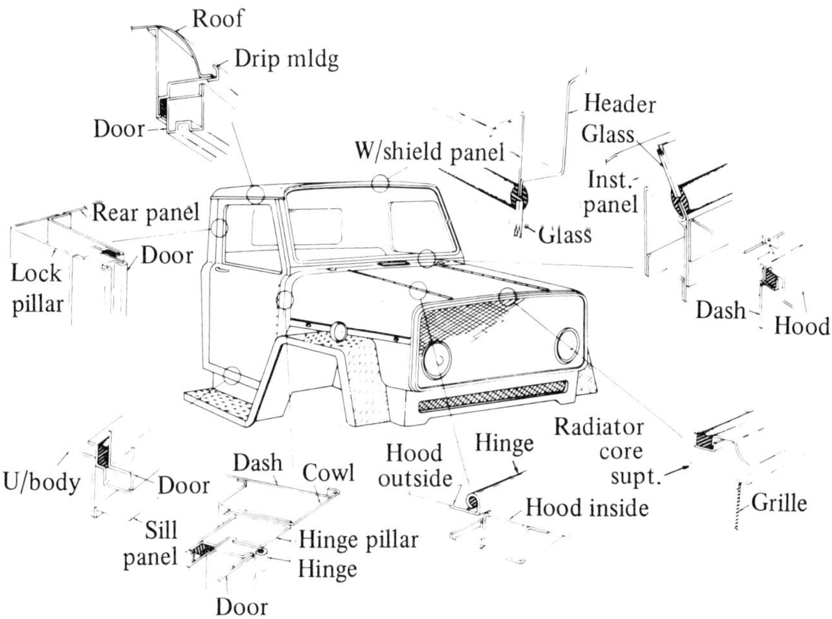

Fig. 6.12. Chrysler XLV truck cab section details

the joints had to be overcome by gussetting; also a number of flat panels had to be reinforced to stabilize them. Generally, however, the squareness resulted in an improved structural performance.

ADHESIVES AND SEALANTS

Structural adhesives are gaining ground in vehicle body production, having been successfully applied to such areas as bonnet panel stiffeners, as a partial replacement for spot welding, and in glass to metal bonding for windscreen fitment. Advantages of adhesive bonding are that an even stress distribution is obtained over the whole of the joint; there is a low melting point and so flammable components can be joined; corrosion is reduced since the adhesives are gap filling; weld marks are eliminated and vibration-absorbing adhesives can be used to damp the motion in thin shell joints.

Main automotive body adhesives are those which can be applied successfully to oily steel surfaces and which will cure in normal paint stoving ovens. The three main categories are thermoplastic, thermosetting, and elastomeric. Of the thermoplastics modified PVC adhesives can exhibit good tensile and shear strength; on stoving they form into solid rubber-like compounds.

Of the thermosets, two epoxy systems do not require heat for curing and a range of impact strengths and flexibilities are available depending on the constituent parts. They do, however, require surface pre-treatment prior to bonding. Also, in the 'in-between' categories are polyurethane compositions which give remarkably high joint strengths but have a toxic hazard. Of the elastomeric group polysulphide rubbers are the most used by vehicle builders, being applied for windscreen adhesion. Perimeters of the screens are prime coated and a bead of polysulphide rubber gunned around the frame to be bonded. After contact and 30 min cold cure the bond is fast and sealed.

Figure 6.13 shows examples of metal jointing: at (a) a reinforcing cross brace for bonnet or boot, at (b) a vee clamp reinforcing a 'T' joint and at (c) a roof panel bonded to its drip rail. The PVC adhesive used in these applications gives a strength of about 800 lbf/in^2 (5·52 MN/m^2) after curing, without the use of jigs or presses during the curing cycle.

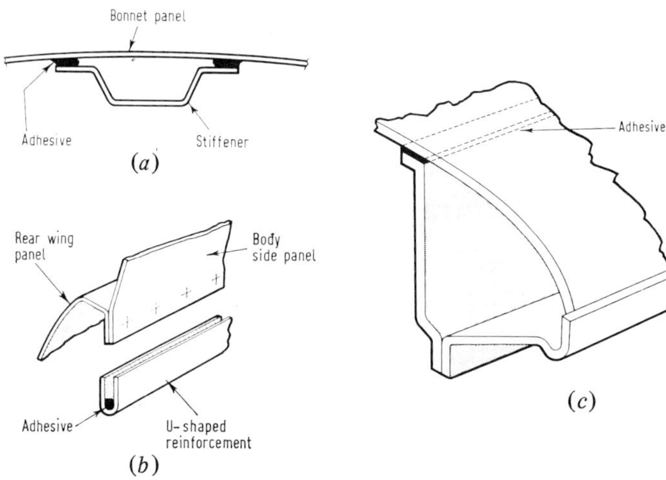

Fig. 6.13. Applications of metal-to-metal bonding

Good joint design is important to achieve the best adhesive performance; peel and cleavage forces should be kept to a minimum. Joint strength falls off outside the temperature range 40–194°F (−40 to +90°C) but increasing the stoving time involves increasing strength with time. In reinforcing sheet structures with flanged stringers, adhesives have an important advantage over spot welds in that the whole of the flange area adheres and stabilizes the sheet, not merely the area of the spot welds.

Sealing of vehicle bodies should also be considered at the design stage; high speed vehicles are particularly susceptible in the high aerodynamic pressures that are built up, particularly so if the window design is such as

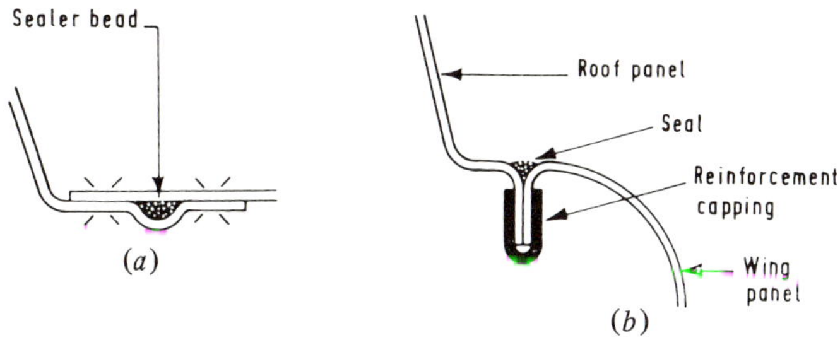

Fig. 6.14. Sealing methods for spot-welded joints

to cause a pressure drop inside the body. Generally, there is a problem in spot welded steel bodies of obtaining a lap joint sealant that does not burn out on welding. One solution is to introduce a bead of sealant between spots as shown in Fig. 6.14 a particularly valuable method in floor pan lap joints. With all sealed joints it is important to design for absolute minimum movement at the joint as cracking of sealant, thus caused, is the commonest form of sealant failure. A-post to sill connection needs special attention in avoiding the passage of water from wheel arch to sill, then subsequently on to the floor pan. Rear wheel arch joints which cannot be accessed for sealing are another common problem, as are joints inside air intake cavities for heaters built in to the dash. Extruded plastisol or paste is usually employed for external joints, Fig. 6.14(b).

GLASS REINFORCED PLASTIC (grp) CONSTRUCTION

For low volume vehicle body manufacture, and for the specialist builder, a greater freedom in body design has come about with the advent of glass reinforced plastic construction. This has also spread to medium rate production cars such as Chevrolet Corvette, Studebaker Avanti, Lotus Elite, and later derivatives. The tooling cost for the Corvette structure was only one tenth of the cost for a comparable steel body. Matched metal dies are employed, to produce 90 bodies per day, working at pressures of 125–400 lbf/in^2 (0·82–2·75 MN/m^2) with the dies heated to 215–300°F (102–149°C). Curing of the resins under these conditions takes 2–5 min.

The glass reinforcement is preformed by spraying chopped fibres and resin binder on to rotating preform screens, Fig. 6.15—the more common mat reinforcement being used only for shallow mouldings with simple curvatures. Powdered or emulsified resin binder is employed and the preform is stoved for 2–3 minutes at about 230°F (110°C) before placing on the mould. A total of only 89 mouldings, considerably less than the equivalent quantity of steel pressings, was used for each Corvette body.

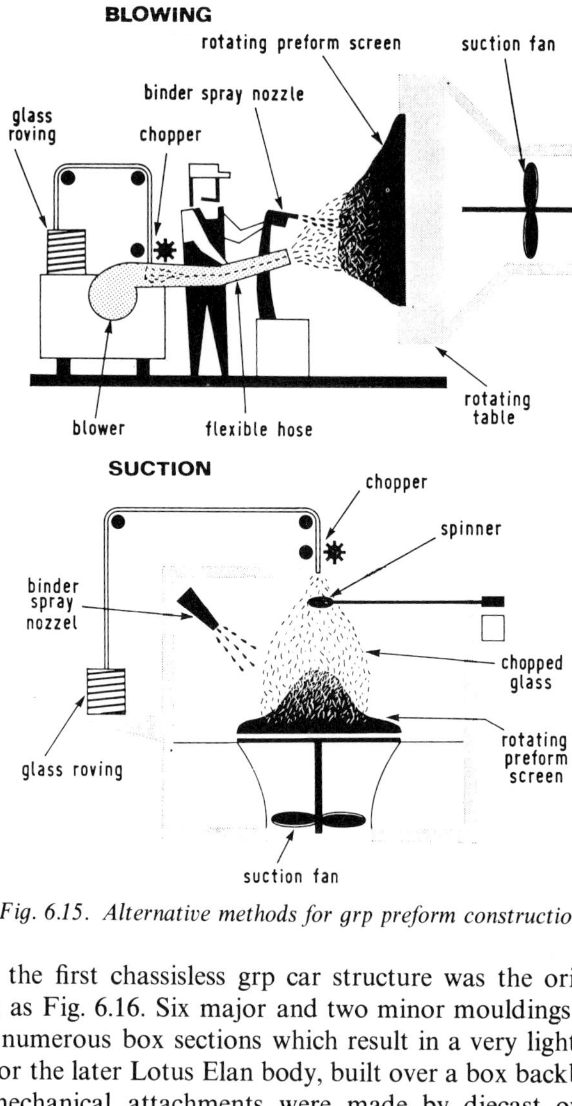

Fig. 6.15. Alternative methods for grp preform construction

Probably the first chassisless grp car structure was the original Lotus Elite shown as Fig. 6.16. Six major and two minor mouldings are bonded up to form numerous box sections which result in a very light and strong structure. For the later Lotus Elan body, built over a box backbone chassis structure, mechanical attachments were made by diecast oval bobbins Fig. 6.17(a) which are fairly easily worked into the design and set into the laminates prior to moulding. They have high resistance to twist and tension and are actually bolted into place on the moulds to ensure correct positioning. The door hinge system is shown in Fig. 6.17(b) and comprises a diecast attachment plate into which is screwed a bolt with a nylon ball at the end. These plates are assembled into the door so that the nylon balls project up through its top surface and down through its lower face. The balls engage in cup bobbins laminated into the adjacent bodywork, result-

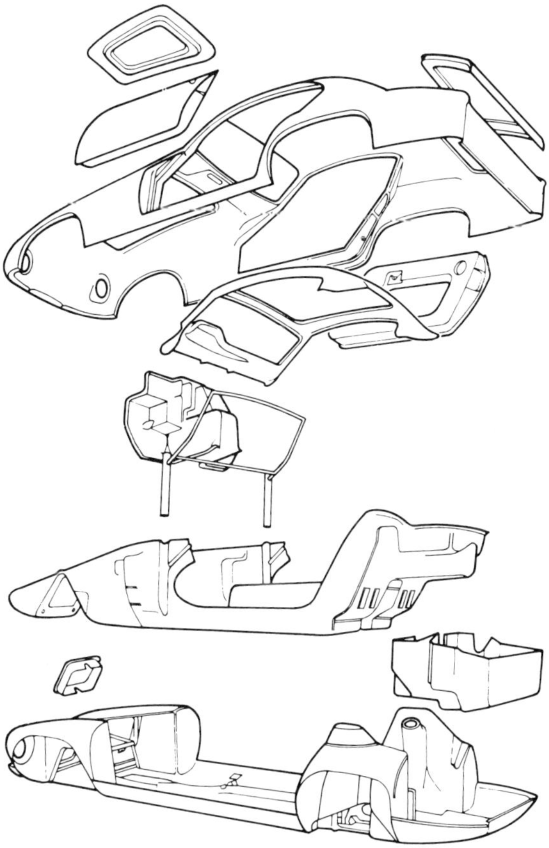

Fig. 6.16. Exploded view of Lotus Elite 1 structure

ing in a wide centred, rigid and free-working door which does not cause stressing of the skin panels.

A typical heavy duty automotive preform grp mix, per cent, is glass fibre 40, resin 40, monomer 0·41, catalyst 0·41, filler 16·50, lubricant 0·08, and preform binder 2. Strength and stiffness properties for such a mix are tensile strength 24 000 lbf/in^2 (165 MN/m^2), elastic modulus $1{\cdot}4 \times 10^6$ lbf/in^2 (9·65 GN/m^2), shear strength 16 000 lbf/in^2 (110·2 MN/m^2), tensile elongation 2 per cent and flexural fatigue strength for 10^7 cycles is 15 per cent of ultimate strength. A 10 per cent improvement in properties can be achieved by raising glass content, to a minimum of 45 per cent, in conjunction with a higher flexibility resin.

Failure in a plastic material occurs at a stress quite distinct from the tensile or compressive stress, namely, the flexural, cross-breaking or bending strength. A laminate also has no yield point, and experience shows that

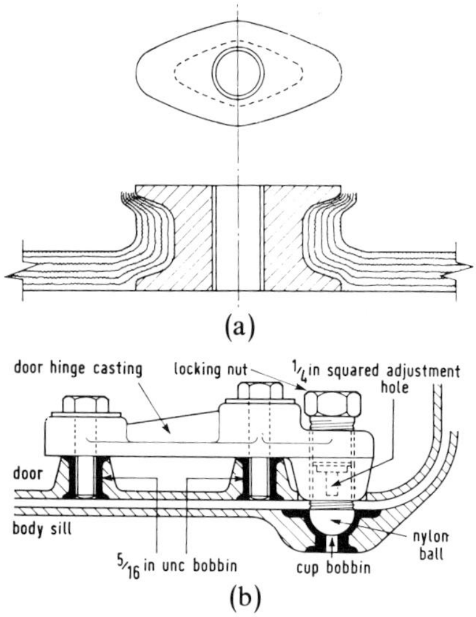

Fig. 6.17. Attachment bobbins for grp structures

minor impacts cause only a spring-back deflection; limiting stresses cause fracture but no permanent set. Hollow sections subject to bending are more likely to fail in tension than to buckle and ultimate tensile strength is the main criterion for specification purposes. Thin walls are desirable in press moulding since they contribute to a reduction in material costs and cure cycle times. Generally thickness t in should be in the range $0{\cdot}03 < t < 0{\cdot}25$ in; above the upper limit a dough moulding is usually more suitable.

Design approach is to integrate parts into a one-piece moulding to save fabrication costs and eliminate sources of stress concentration. Use double curvature to increase stiffness and use well-radiused curves to assist fibre distribution and air removal. Allow as much taper as possible in deep draw mouldings, avoiding any undercuts, to facilitate ease of removal from the mould. Mould-in ribs, hollow sections, and lightweight cores so as to increase local strength and stiffness. Since stress concentrations are not relieved by yielding, adopt large area washers and metal inserts to spread shear and bending loads at attachment points. Adopt the following minimum safety factors:

Short term static	2	Repeated	5
Long term static	4	Reversed alternating	5
Variable unidirectional	4	Impact loads	10

Fatigue of grp

The fatigue life of grp has been considered by M. J. Owen in an article in *OEM Design* January 1972. He explained that earlier researchers had established the fatigue strength of grp at 10^6 cycles under fully reversed stress conditions as lying between 0·20 and 0·35 times the mean static strength. Attention had also been paid to obtaining constant life (Goodman) diagrams of mean stress versus stress amplitude, fatigue properties of grp being shown to be sensitive to mean tensile stress. The experimental mean stress amplitude usually fell below the Goodman line used as a predicter for structural metals; see Fig. 6.18.

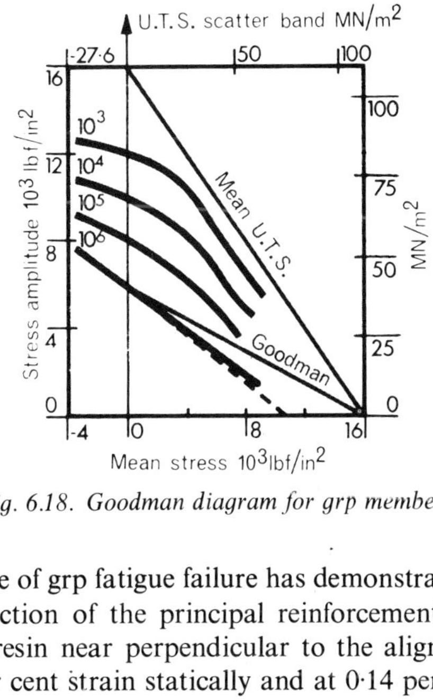

Fig. 6.18. Goodman diagram for grp member

Study of the nature of grp fatigue failure has demonstrated that laminates stressed in the direction of the principal reinforcement suffer separation between fibre and resin near perpendicular to the aligned fibres, damage occurring at 0·3 per cent strain statically and at 0·14 per cent strain under 10^6 cycles fully reversed stress. Progressive damage occurring is shown in the S–N diagram of Fig. 6.19 and the designer must choose a failure criterion to suit the application.

Sandwich panel construction

A method of enhancing the stiffness of grp panels is the use of sandwich panels made possible by reservoir moulding. P. R. Chant of Shell read a paper at the Eighth International Reinforced Plastics Conference (1972) explaining the process. Its basic feature is the use of a reservoir of flexible open-cell foam impregnated with resin which is sandwiched between two

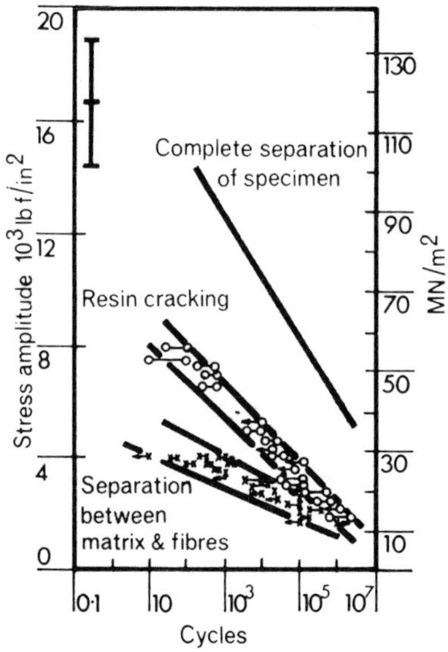

Fig. 6.19. S–N curves for grp showing progressive damage

layers of dry fibre reinforcement. The three layer sandwich is placed between male and female dies which are brought together to provide a moulding pressure of only 14·2–28·4 lbf/in^2 (1–2 kg/cm^2).

The foam acts as a sponge and during the moulding operation the resin is squeezed from it to impregnate the fibre. With the reinforcement concentrated at the surfaces it contributes most to the flexural stiffness; the stiffness of the sandwich is varied by applying different degrees of compression in the mould, see Fig. 6.20. A variation on the process is to use a pre-formed foam reservoir with a two part mould comprising one rigid and one flexible member, such as a liquid-filled bag exerting constant pressure.

A sandwich moulding technique is also applicable to thermoplastics and has been described by ICI researchers. Two different polymers are injected sequentially into a mould and the size of the cavity in which the moulding is being introduced is increased by moving one half of the mould a short distance, to allow the foaming agent in the core material to expand. One of the largest sandwich components yet made in this way is a bonnet for the BLMC 1100 car, measuring 3·7 × 3·0 ft (1·12 × 0·91 m). It weighs 9·9 lb (4·5 kg) against 22 lb (10 kg) for the metal component and it has polypropylene skins and foamed polypropylene core. Glass or asbestos reinforcement of the skins is also possible to raise further the specific stiffness.

Visco-elastic analysis

When designing in thermoplastics the gradual deformation under continuous load, known as creep, must be taken into account. Strain cannot be

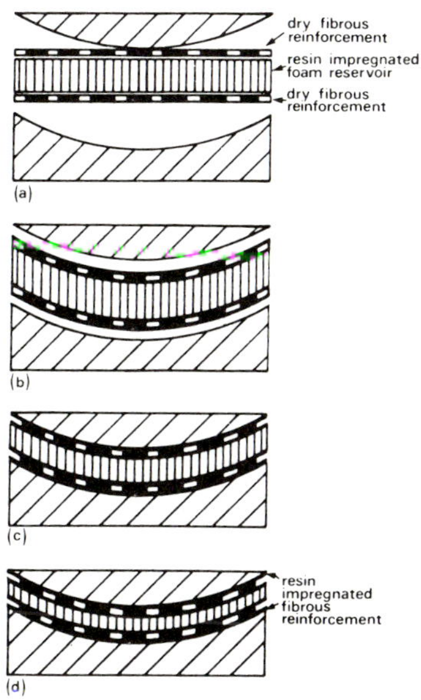

Fig. 6.20. Reservoir moulding of grp sandwich panel

represented as a single valued function of stress except as an approximation of the real behaviour when the duration of loading is short. For small strains, however, certain plastics can be regarded as linearly viscoelastic; for example, flexural deflection of a weightless beam length L under central load W is $= WL^3/48E_t I$ where E_t is the creep modulus which varies with load duration. E_t may be chosen for a particular deflection by means of successive approximation. Figure 6.21 shows that this is proven for the case of a sandwich beam for loads up to 350 hr duration, where experimental and theoretical results closely coincided.

Pseudo-elastic design, so called, relies first on a completed elastic design of the structure by normal methods. Thereafter suitable values of elastic

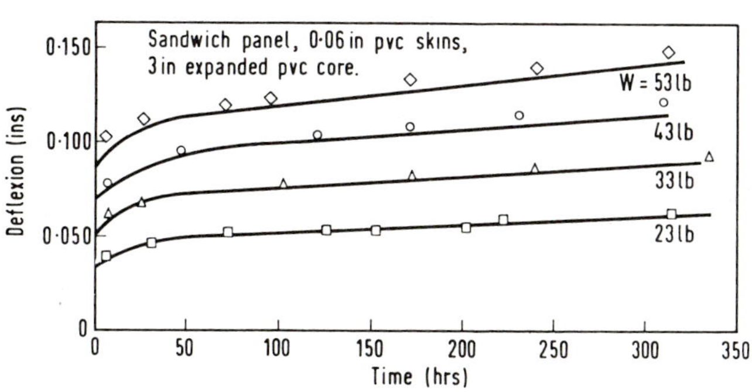

Fig. 6.21. Prediction of creep modulus in thermoplastic sandwich

modulus E are required. The design approach is first to establish component function and service conditions in terms of expected lifetime and maximum in-service temperature. Next the worst case is considered, of the component operating continuously at the maximum service temperature under a constant applied load. Then the value of tensile creep modulus is chosen appropriate to the maximum strain, service temperature, and chosen lifetime from manufacturers' tables. The deformation formula is chosen from the appropriate classic analysis and the value of working stress or modulus is inserted to find the deflection.

For example, an extruded T-section strip in acetal copolymer, 1 in (2·54 cm) wide, 1 in (2·54 cm) deep with general wall thickness 0·2 in (5·08 mm) is considered. This is to be used as a 5 ft (1·52 m) long encastré beam which will operate under self-weight conditions at 20°C for 10 yr.

Section second moment of area $I = 0{\cdot}0313\ \text{in}^4\ (1{\cdot}304\ \text{cm}^4)$

Cross-section area $= 0{\cdot}36\ \text{in}^2\ (2{\cdot}32\ \text{cm}^2)$

Density of acetal copolymer $= 0{\cdot}51\ \text{lb/in}^3\ (14{\cdot}12\ \text{g/cm}^3)$

Assuming a low strain, less than 1 per cent, then the estimated tensile creep modulus from tables is $1{\cdot}1 \times 10^5\ \text{lbf/in}^2$ at 1 per cent strain, 10 yr, 20°C. Maximum deflection is given by

$$y = \frac{1}{384}\frac{Pl^3}{EI} = \frac{\rho AL^4}{384EI} = \frac{0{\cdot}51 \times 0{\cdot}36 \times (60)^4}{384 \times 1{\cdot}1 \times 0{\cdot}0313 \times 10^5} \quad \text{after ten years.}$$

$$= 0{\cdot}18\ \text{in}\ (4{\cdot}56\ \text{mm}) \quad \text{for force } P\text{, density } \rho \text{ and flexural stiffness } EI$$

Thermoforming

'Solid skin' thermoplastic bodies have also been built both experimentally, and for volume production in the case of the Citroen Mehari. ABS is the material used, as for example, in the Cycolac research vehicle shown in Fig. 6.22 made from two sheets of 6 ft × 4 ft (1·825 m × 1·22 m) × $\frac{1}{4}$ in (6·35 mm) thick flat ABS extrusions, unreinforced. Welded tubular steel subframes carry suspensions and power transmission unit. The sheets are separately vacuum-formed into mating upper and lower body halves which are bolted together along the waistline flange.

Infrared radiant heaters are used to soften the sheets before closing the aluminium filled epoxy resin moulds and applying suction. Total depth of draw is 30 in (0·762 m) which is the criterion of material thickness rather than the working stress in the finished surface. Moulding time for each half of the body is 15 min, after which the only finishing required is edge trimming. The resulting car with Hillman Imp running units weighed 960 lb (435 kg).

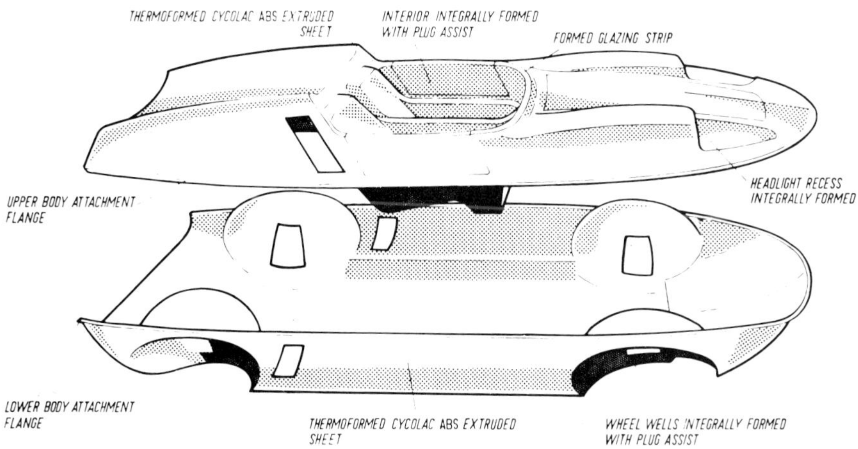

Fig. 6.22. Cycolac research vehicle structure

It is not possible to produce thermoformed parts with abrupt changes in wall thickness and it is difficult to obtain sharp detail on the opposite side of the mould. The degree of taper can vary with depth of draw but generally a draft of 2–3 degrees is the minimum for male moulds. Female moulds are much more tolerant because of the tendency of the formed part to shrink away from the mould and here a draft angle of only 1 degree is adequate.

In general radii should always be as generous as possible; the smallest radius which can be used will vary with wall thickness. Particular areas of a thermoforming can be thickened up by local cooling of the sheet prior to forming, and zoned heaters can be used with similar effect during the heating period. Figure 6.23 shows the effect of pre-stretching the heated sheet with compressed air so that the male mould can be moved into the bubble before forming. This method considerably improves wall thickness distribution.

It is possible to join thermoplastic components by ultrasonic welding, in

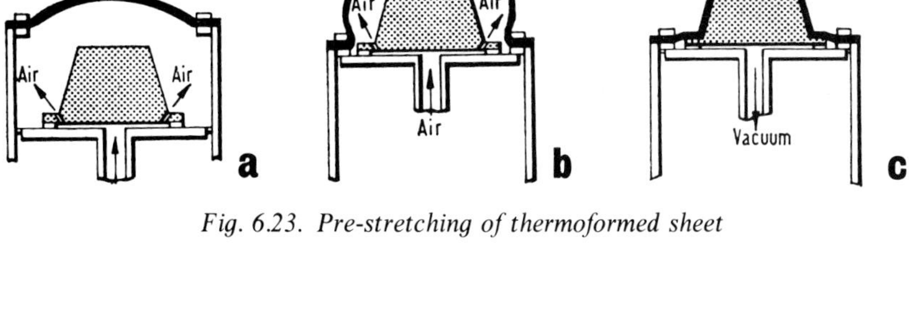

Fig. 6.23. Pre-stretching of thermoformed sheet

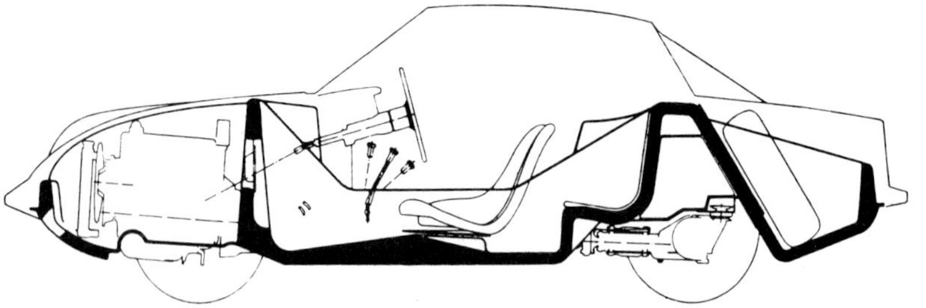

Fig. 6.24. Bayer experimental car

which high frequency vibrations (around 20 000 Hz) excite the parts being joined; this generates heat through friction, melting adjacent plastics to form the weld.

Combined thermoplastic and thermosetting materials have also been used in vehicle body construction. A notable example is the Bayer experimental car shown in Fig. 6.24. Here a grp skinned polyurethane cored sandwich panel is employed. The edges of the grp skins are cemented to form a cavity and liquid polyurethane introduced. Rising foam then adheres strongly to the faces and sets in a few minutes. Thirty minutes lapse is necessary before moving the sandwich panel from the mould. Metallic inserts cater for engine and running gear attachment points. The polyurethane core density nominally of 1 g/cm^3 has a distribution across the core thickness from 0·4 at the centre to 1·2 g/cm^3 at the skins. Tests have shown at the running gear anchorages that shear forces of 1 lbf/ft (150 kg/cm) and tensile forces of 660 lbf (2·94 kN) have been withstood by the anchorage bolts. Fatigue tests have shown a permissible alternating flexural stress of 7800 lbf/in^2 (375 kN/m^2) for the sandwich panel.

CHAPTER 7

GOODS VEHICLE STRUCTURE DESIGN

FROM a structural viewpoint the goods vehicle divides into three major categories: the integral panel van, the separate-chassised load carrier with platform, box van or special purpose body, and the articulated vehicle with trailer. Car-derivative and panel vans are in the main considered by structural analysis methods akin to those used to study car structures.

The load-carrying haulage truck has a chassis frame upon which a body is mounted which contributes to the rigidity of the whole vehicle. Beam axles located by laminated leaf springs ensure accurate directional stability even under twist of the chassis frame. Thus an open platform bodied truck for which the body contributes minimum stiffness can flex in operation.

CHASSIS FRAME CONFIGURATION

Ladder-type frames have been the almost universal form taken by truck chassis since their inception and show no signs of being superseded in the immediate future. Bending strength is built into the side members, while the cross members stabilize the side members against local twisting due to offset loading as well as supporting the engine/transmission system. In the case of lighter vehicles such as panel vans and pick-ups, using independent front suspension, torsionally stiff cruciform frames are usually employed.

Until soft truck suspensions are developed with sufficient vertical travel to minimize distortion of the 'base' upon which the body is mounted, there is a requirement for the ladder frame of the medium and heavy truck to 'weave' freely under arduous road and site conditions, necessitating a torsionally flexible frame. This is particularly so with tipper chassis and four-wheel-drive cross-country vehicles. Side members are normally of channel section, with the web supporting most of the downward thrust of body and payload; top and bottom flanges give lateral support to the web and provide bending strength to the side member. As web depth is increased and flanges move farther apart, section modulus is increased with a corresponding increase in carrying capacity. For specialist vehicle applications, and for local reinforcement of bending strength, frame flitching is employed with composite sections made from added L-section or enveloping channel members.

Body mountings are designed to feed loads into the webs of the channel side members and in some cases mountings are designed so that the line of action of the applied load passes through the shear centre of the section. This ensures that there is no local twisting of the section.

The overall functions of the frame are to carry the body and payload, to retain alignment of the axles and driveline and to transmit the steering, driving and braking forces from the respective axles to the mass of the vehicle. Four general modes of loading are vertical bending and torsion—mentioned above—and axial and lateral bending, the last two causing parallelogramming and lateral deflection, twist, or local warping of individual side members.

Method of attachment of cross members to side members, by flange or web joint, varies between manufacturers. Generally the lighter high-volume vehicles employ the flange method, while the reverse is true of the heavier low-volume models. Spacing of cross members is close enough to avoid side member failure by buckling of the compression side. Some diagonal bracing must also be supplied by the cross members to provide resistance to lozenging.

The chassis frame must be sufficiently flexible torsionally to accept the required deflections without being overstressed, a question of resilience rather than strength. Conventional frame members are essentially quite flexible torsionally, being of open section, but as they are short their resilience is reduced by the effect of end-restraint at the attachments. Use of more flexible end connections is not the normal solution, however, particularly if the cross member has to act as a cantilever beam outboard of the side members, as at spring brackets. The solution is to maintain adequate beam strength while modifying the cross member section to secure more torsional flexibility. This combination of strength and rigidity at the attachments and flexibility in the long spans is obtained in the so-called 'alligator' cross member.

With the gradual improvement in ride comfort of trucks, manufacturers are beginning to consider softer suspensions, with the attendant need to re-examine the stiffness of the chassis frame. The structural flexing of the frame also needs to be examined during design with the objective of building in maximum fatigue resistance.

Side member torsion

The conventional frame has considerable stiffness in symmetrical bending but less in torsion (asymmetrical bending). The torsionally flexible frame thus has open section side members and often open section cross members. Such resistance to twist as is present relies on the differential bending, as opposed to pure torsion, of the frame member. In pure torsion the members do contribute to the overall twist resistance, the extent of which depends on the partial axial warping of the members. This limitation is well documented by Tidbury *et al.* in the 1968 FISITA paper referred to later in this chapter. The concept of differential bending can best be understood by visualizing a thin card model of a short length open section beam such as the I beam in Fig. 3.14. Notice that with ends free, except for transverse torsional reaction, the beam develops its torsional stiffness by differential bending of the flanges in their own planes. Clearly if the ends were encastré such that displacement along the axis of the beam is totally restrained, there will be much greater torsional rigidity, arising from the additional end-fixing-moments in the plane of the flanges. Figure 7.1 shows the decrease in differential bending torque as the axial warping restraint is progressively less towards the end of the beam shown in Fig. 3.14 from the centre. Simultaneously the 'St Venant' torque, measured by the Bredt Batho relationships in an earlier chapter, increases from zero to a partial contribution to the torsional resistance at the ends as the differential torque diminishes. Figure 7.1 should be read in conjunction with Fig. 3.14,

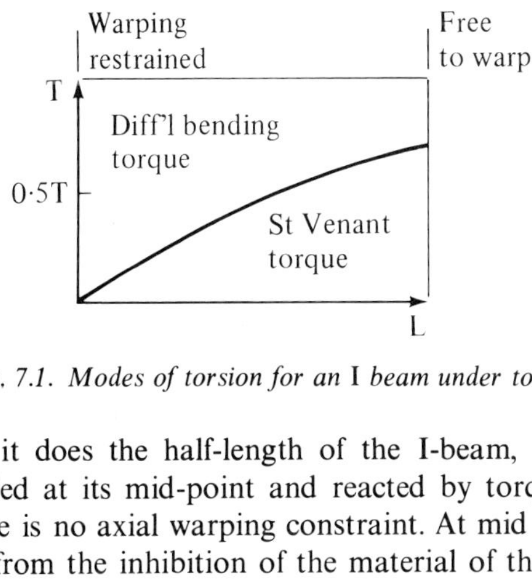

Fig. 7.1. Modes of torsion for an I *beam under torque*

representing as it does the half-length of the I-beam, length $2L$, under torque $2T$ applied at its mid-point and reacted by torques T at the free ends where there is no axial warping constraint. At mid span full warping restraint arises from the inhibition of the material of the beam extending out to its two ends. For the semi-trailer chassis frame in particular torsional stiffness is required in the frame for such manoeuvres as accelerating the tractor at an angle to the semi-trailer when driving out of a corner. In such a situation the torsional stiffness of the frame is called upon to react the inertia torque due to the height of the payload centre of gravity above the level of the coupling turntable.

Chassis frame analysis

This point is made by T. H. G. Megson writing in *Automotive Design Engineering* May 1970†; he also provides the following analysis of frame

† See comment by Guy Tidbury, p. 151.

torsional stiffness. It is based on the simplifying assumption that the torsional stiffness of the chassis frame is the sum of the torsional stiffnesses of its members. For closed tube cross members torsional rigidity

$$G \,.\, J_{\text{closed}} = G\frac{4a^2b^2t}{2(a+b)}$$

for a hollow rectangular section member depth $2a$, breadth $2b$, and thickness t. If the thin tube is slit along one side to form an open section

$$G \,.\, J_{\text{open}} = G\frac{2(a+b)\,.\,t^3}{3}$$

For $a = 6$ in (152·3 mm), $b = 2$ in (51 mm), $t = 0{\cdot}1$ in (2·54 mm) then $GJ_{\text{open}} = 0{\cdot}0053G$ and $GJ_{\text{closed}} = G \times 3{\cdot}6$ lbf/in^2 (24·0 kN/m^2), a ratio of 680 : 1.

These are St Venant torsional stiffnesses and neglect the effect of differential bending. In this analysis it is assumed that the members are rigid in bending and the deflection mode is considered to be as shown in Figs 7.2 and 7.3 with axes of the plane sections remaining plane. The analysis is based on the use of flat plate cross members and channel section side members. Flat plates were assumed to eliminate the effect of axial constraint on the warping of the cross section which is difficult to calculate unless the degree of rigidity of the connection is known.

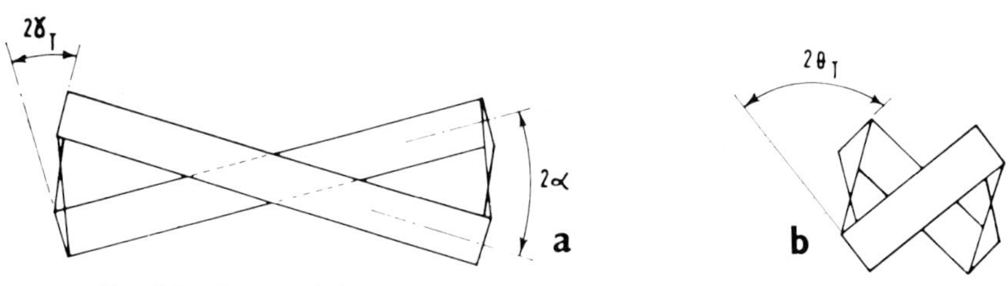

Fig. 7.2. Torsional deflection of single bay of idealized chassis frame

From Fig. 7.3 applied torque to the whole chassis frame

$$T = 2G_bJ_b\frac{d\theta}{dx} + n\,.\,G_cJ_c\frac{d\gamma}{dz}$$

and for small deflections: $v = b\theta$ and $\alpha = v/z = b\theta/z$ and for rigid connections between cross member and side member beams $\gamma_t = \alpha$ and $\gamma = (x/b)\alpha$ so $d\gamma/dx = d\theta/dz$ and $T = (2G_bJ_b + n\,.\,G_cJ_c)\,d\theta/dz$.

Guy Tidbury of Cranfield School of Automotive Studies commented† on Mr Megson's results verifying the expression obtained, given complete

† See previous footnote.

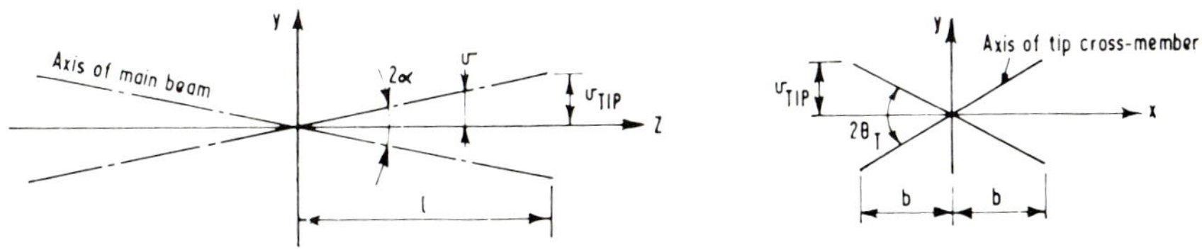

Fig. 7.3. Longitudinal and transverse twist of single bay

warping inhibition at the joints, and largely warpless cross members sectioned. It was pointed out, however, that the warping inhibition was only justified for heavily gussetted frames and that conventional chassis frames required a more thorough analysis allowing for partial warping inhibition at the joints and typical member cross sections. In an analysis by Guy Tidbury and co-workers discussed at the 1968 FISITA congress the concept of partial warping is examined in detail. Depending on the extent of the fixing between the cross members and side members, the end constraint provides a variable degree of axial warping in the cross members under torsion. The paper describes alternative methods for estimating somewhat idealized connections of channel cross members, butt welded to the webs of channel side members, and the more realistic case of the alligator-type cross member. Practical tests are described on frames having different forms of cross member type and end fixing. Table 7.1 shows the torsional stiffness in lbf ft/degree for these frames using 'free', 'inhibited' and 'partial' warping theory. The effectiveness of the 'partial warping' theory is clearly demonstrated in comparison with the experimental results obtained. In the paper an expression was provided due originally to Erz for a truck frame, wheelbase e, distance spring hangers d, effective length of cross members d_1

TABLE 7.1

	Torsional stiffness lb ft/degree			
	Theoretical			*Experimental*
	Free warping	*Inhibited warping*	*Partial warping*	
Frame No. 1a	18·0	255	70·3	62·5
Frame No. 1b	18·0	255	88·5	67·2
Frame No. 1c	18·0	255	103·5	114·5
Frame No. 2a	58·6	750	192·5	132·0
Frame No. 2b	58·6	750	226	174·9
Frame No. 2c	58·6	750	251·5	188·5
Frame No. 3	—	—	132	153

and frame width between side member torsional centre d_0. Torsional stiffness between axle centrelines

$$S = \frac{G \,.\, d^2}{d_l e^2}\left(\sum_{n=1}^{n=n} J_{c,n} + \frac{2d_1}{d_0^2}\sum_{m=1}^{m=n-1} J_{b,n} \,.\, l\right)$$

where l is the length of bay and G is the shear modulus.

Torsion constants, J, for thin-walled sections are given alongside second moments of area, I_{zz} in Figs. 7.4 and 7.5.

Stiffness of added cruciform bracing, Fig. 7.6, can be estimated by considering the cruciform as two beams with fixed ends, each imposing load on the other at central span. Centre deflection on each beam

$$d = \frac{W \,.\, l_c^3}{192EI_c} = \frac{2F(l_1^2 + l_2^2)\sqrt{(l_1^2 + l_2^2)}}{192EI_c}$$

Thus torque for unit angular deflection is

$$\frac{F \,.\, l_2^2}{4d} = \frac{24 \,.\, E \,.\, I_c \,.\, l_2^2}{(l_1^2 + l_2^2)\sqrt{(l_1^2 + l_2^2)}}$$

For this simplified example constant section beams form the bay diagonals making up the brace. In practice higher specific rigidities are obtainable with tapered (constant stress) beams. In all cases resulting corner deflection is four times the centre deflection of each beam.

Stressing of frames is carried out by combining bending stresses and shear stresses in the frame side member as a principal combined stress given by

$$\sigma_{max} = \frac{\sigma_b}{2} \pm \sqrt{\left\{\left(\frac{\sigma_b}{2}\right)^2 + \tau^2\right\}}$$

In a straight side member, bending and shear stresses are the only components of the principal stress. In a curved beam the transverse bending and radial stress must also be considered. As shown in Chapter 3 bending and shear stress are given by

$$\sigma_b = \frac{M \,.\, y}{I} \quad \text{and} \quad \tau = \frac{S \,.\, A \,.\, \bar{y}}{I \,.\, b}$$

For bending moment M, shear force S, y is the distance from neutral axis to extreme fibre, I is the section second moment of area, b is the breadth, $\bar{y}$ is the distance from neutral axis to centroid of area above shear plane. Bending deflections of side members are normally derived graphically by integration of bending moment diagrams.

Sometimes, however, chassis frames fail by fatigue cracking of the compression flange of channel section members, attributed to compressive

OPEN SECTIONS

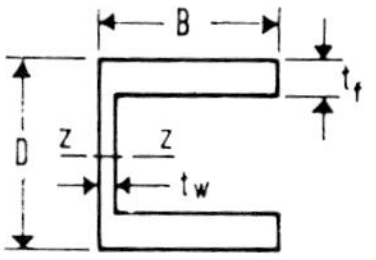

$$`J' = \frac{2\,Bt_f^{\,3} + (D - 2t_f)t_w^{\,3}}{3}$$

$$`I_{ZZ}' = \frac{BD^3 - (B - t_w)(D - 2t_f)^3}{12}$$

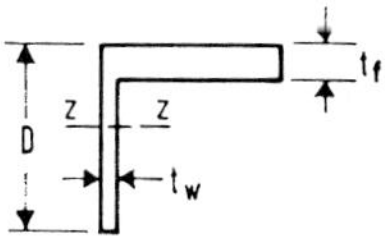

$$`J' = \frac{B\,.\,t_f^{\,3} + (D - t_f)t_w^{\,3}}{3}$$

$$`I_{ZZ}' = \frac{B\,.\,D^3 - (B - t_w)(D - t_f)^3}{12}$$

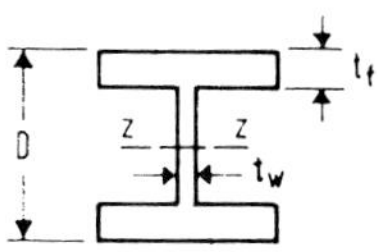

$`J' =$ } as for channel above
$`I_{ZZ}' =$ }

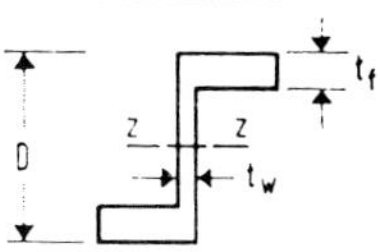

$`J' =$ } as for channel above
$`I_{ZZ}' =$ }

Fig. 7.4. Properties in torsion and bending of open-section members

CLOSED SECTIONS

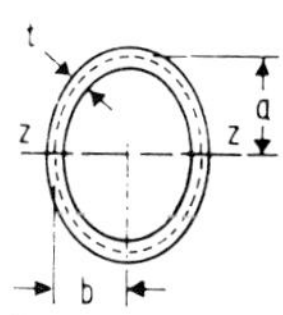

$$`J' = \frac{4\pi a^2 b^2 t}{\sqrt{2(a^2 + b^2)}}$$

$$`I_{ZZ}' = \frac{\pi}{4}\left\{\left(a + \frac{t}{2}\right)^3\left(b + \frac{t}{2}\right) - \left(a - \frac{t}{2}\right)^3\left(b - \frac{t}{2}\right)\right\}$$

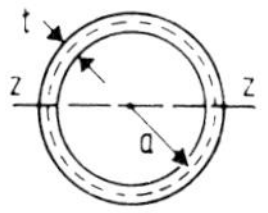

$$`J' = 2\pi a^3 t$$

$$`I_{ZZ}' = \frac{\pi}{64}\left\{(2a + t)^4 - (2a - t)^4\right\}$$

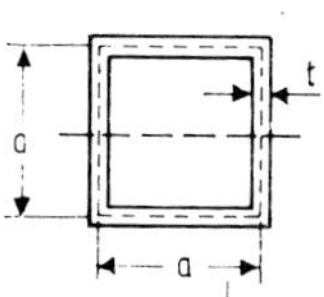

$$`J' = a^3 t$$

$$`I_{ZZ}' = \frac{(a + t)(b + t)^3 - (a - t)(b - t)^3}{12}$$

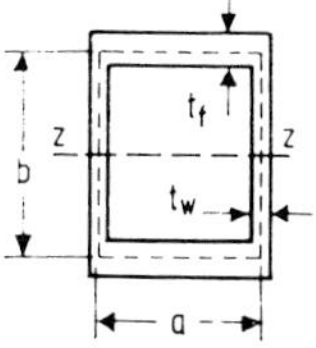

$$`J' = \frac{2a^2 b^2 t_w t_f}{bt_w + at_f}$$

$$`I_{ZZ}' = \frac{(a + t_w)(b + t_f)^3 - (a - t_w)(b - t_f)^3}{12}$$

Fig. 7.5. Properties in torsion and bending of closed-section members

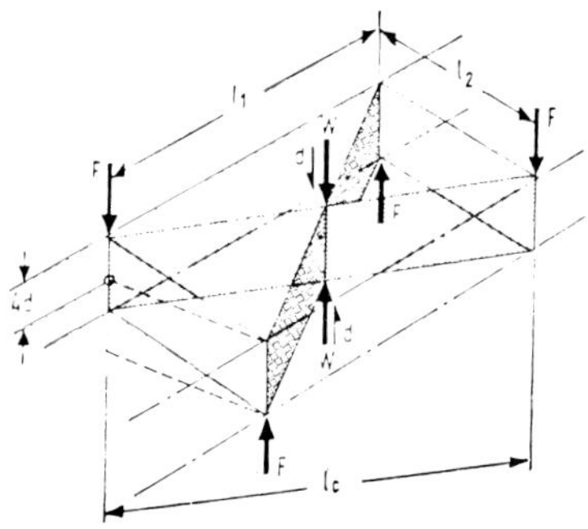

Fig. 7.6. Cruciform bracing of a chassis frame

buckling of wide unstable flanges. This can be aggravated by the use of longitudinal body bearers sitting directly on the chassis terminating abruptly at an area of high stress, perhaps immediately behind the cab, leaving the compression flange suddenly unstabilized. Recent investigations suggest that this form of cracking may be attributable to 'Bimoment' effect at the ends of the body bearers. In consideration of the axial warping of open section members, V. Z. Vlasow, writing on 'Thin-walled elastic beams' in the *Israel Programme for Scientific Translations* (1961), defines the quantity 'Bimoment' having dimensions of moment × length and best considered as a pair of opposite moments M separated by a distance h.

While it is well known that twisting of thin tubes causes some end warping, and subsequently axial loading, conversely end loading of a thin-walled beam will generally cause some torsion and additional local stress (unless a section is 'warpless' by virtue of its geometry). If a bending moment is applied in a plane parallel to the longitudinal axis it can be resolved into a moment, in a parallel plane through the shear centre, and a bimoment. (This is somewhat analogous to the well-known concept of a transverse force acting at a point along the length of a beam, some distance from a reference point, being equivalent to the some vertical force at the reference point plus a pure couple of magnitude equal to that of the force times the distance of the application point to the reference point.)

In the case of an open channel section, equal and opposite forces P, acting a distance d apart on the flanges, represent an applied bending moment Pd and in turn results in an additional bimoment Pdh (where h is the distance across the section to the shear centre), as well as a shear force P. This induces additional flange end loads to those caused by the ones developed in pure bending and could cause compression buckling.

Some manufacturers advise against longitudinal bearers in favour of individual outrigger brackets. These are arranged so that the load is applied at the shear centre of the channel side member. For the arrange-

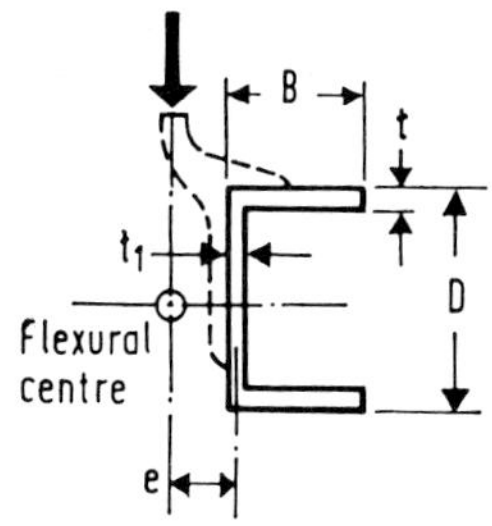

Fig. 7.7. The flexural centre of a channel section

ment shown in Fig. 7.7, the distance *e* required for the line of action of vertical load to give no twist of the channel section is given by

$$e = \frac{B\,.\,t}{4I}(B - t_1)(D - t)(D - 2t)$$

where *I* is the section second moment of area.

STRUCTURAL PURPOSE OF CHASSIS FRAME MEMBERS

William Sidelko writing in SAE paper SP 276 examines the fundamentals of chassis frame design and points out that the legal restrictions on length, axle spacing, and axle loadings forms the first guide to fixing frame dimensions. SAE recommended practice sets frame width at 34 in and in the case of the popular articulated vehicle tractor, coupling turntable height is standardized for different tyre sizes which fixes an 'above frame' dimension. Suspension movement clearance requirements then fixes a 'below frame' dimension that generally results in a maximum frame depth requirement of 10 in.

Four general loading modes listed by the author are vertical bending, torsion, lateral bending and twist, or local warping of individual side members. The latter is mainly due to cantilevered component mountings such as engines, fuel tanks, batteries, and suspension attachments. The effect of twist is overcome with proper cross member placement. Lateral bending can be caused by reactive steering force in overcoming the natural tendency for tandem rear axles to resist turning manoeuvre, for example. The following behaviour characteristics are listed as of prime interest: deflection of the structure at different loads, intensity of loading that will produce the first appreciable permanent set or yielding, intensity of loading that will produce the first appreciable buckling deformation and the effect of cyclic loads on fatigue strength.

First step in frame analysis is treatment of the side member by simple beam theory whence stress σ is given in terms of bending moment M and section modulus Z in $\sigma = M/Z$. Allowable stress is taken as one third of yield stress arising from a dynamic factor of 1·75 and a general 'safety' factor of 1·70. Figure 7.8 shows a series of maximum static bending

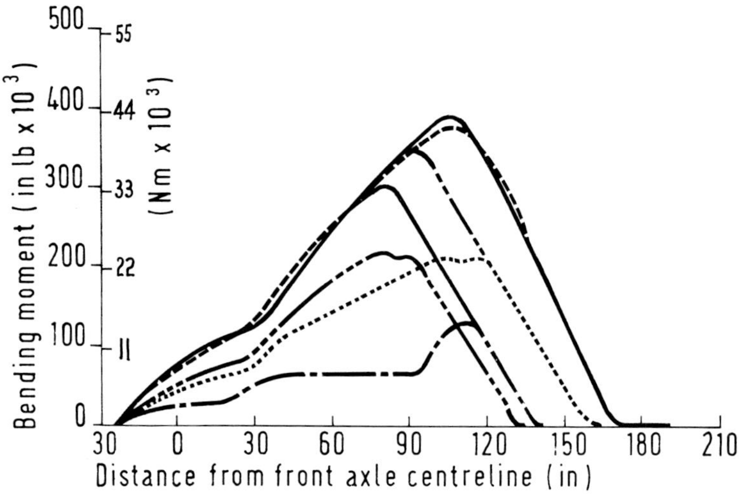

Fig. 7.8. Bending moment plots for a tractor chassis range

moment plots for a range of highway tractors from which maximum bending moment is 396 000 in lbf (44 700 Nm). Choosing a steel of 110 000 lbf/in^2 (758 MN/m^2) yield stress, allowable stress becomes 36 700 lbf/in^2 (248 MN/m^2) and section modulus 10·8 in^3 (193 cm). Section properties can then be obtained from standard formulae.

In studying frame deflection the tractor chassis can be idealized as in Fig. 7.9. Maximum deflection is given by

$$\Delta_{max} = \frac{Fk(l-k)^3\{\frac{1}{3} + \frac{2}{3}(l/l-k)\}^{3/2}}{3EIl}$$

For a 142 in (3·6 m) wheelbase, 24 in (0·51 m) kingpin offset, and allowable frame deflection of 0·4 in (10·2 mm), second moment of area required is 51 in^4 (2120 cm), for example. Standard formulae can then be used to obtain section properties. The criterion to be considered is flange buckling for which the stability curve in Fig. 7.10 is given in terms of critical compressive stress and flange width to thickness ratio, from which 'safe' sections can be checked. It should be noted, however, in this analysis the author appears to ignore warping restraint stress.

Frame cross members need next to be examined in the light of their

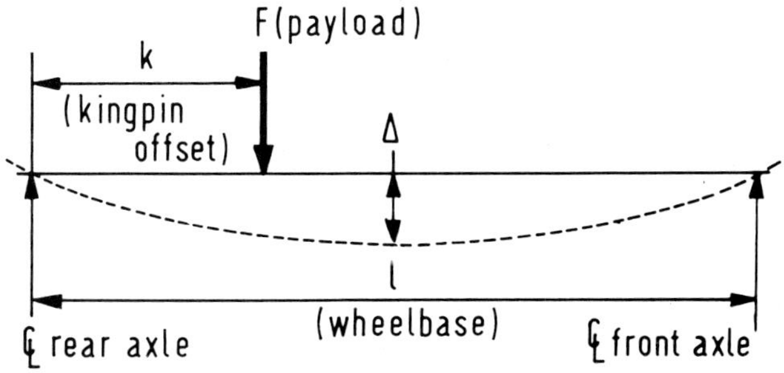

Fig. 7.9. Idealization of tractor chassis in bending

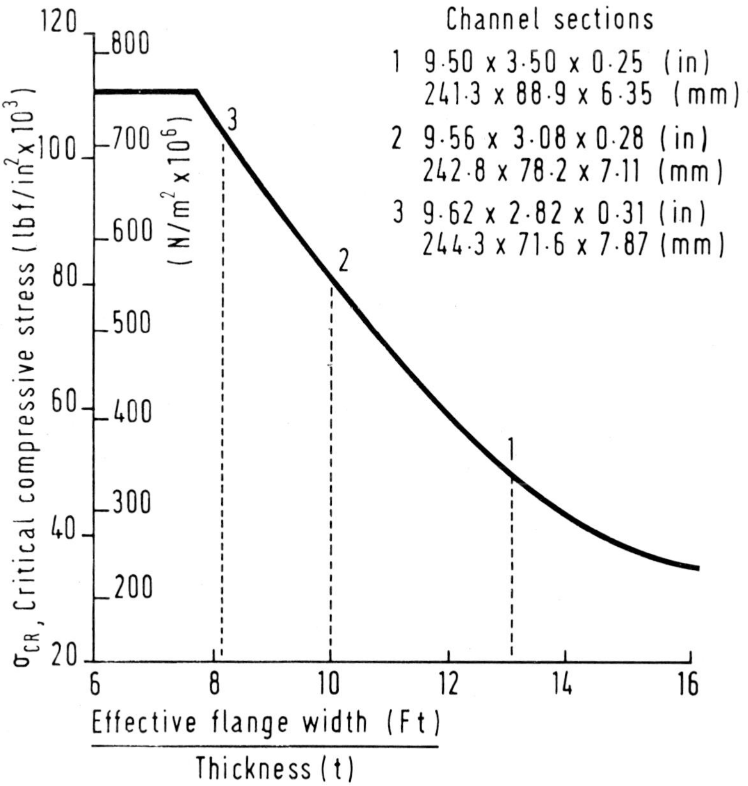

Fig. 7.10. Flange buckling stability curve

fundamental requirement to provide support for installation of vehicle components, separate the side members, and sustain the dimensional integrity of the frame structure and, finally, provide resistance to twisting, torsion, and parallelogram forces applied to the frame. In the case of tandem rear suspensions, cross members must also provide a rigid platform for reacting dynamic load inputs caused by articulation of the axles.

SECTION PROPERTIES BY COMPUTER

Logical selection of chassis side member parameters to suit a range of trucks depends on a number of factors which can be most conveniently examined by computer. G. C. Carver in SAE paper 690 174 has provided a rational approach to side member design based on graphical representation of computed section properties.

Start point is calculation of bending moments from a diagram of principal loads on the side member and the axle load reactions. He chooses

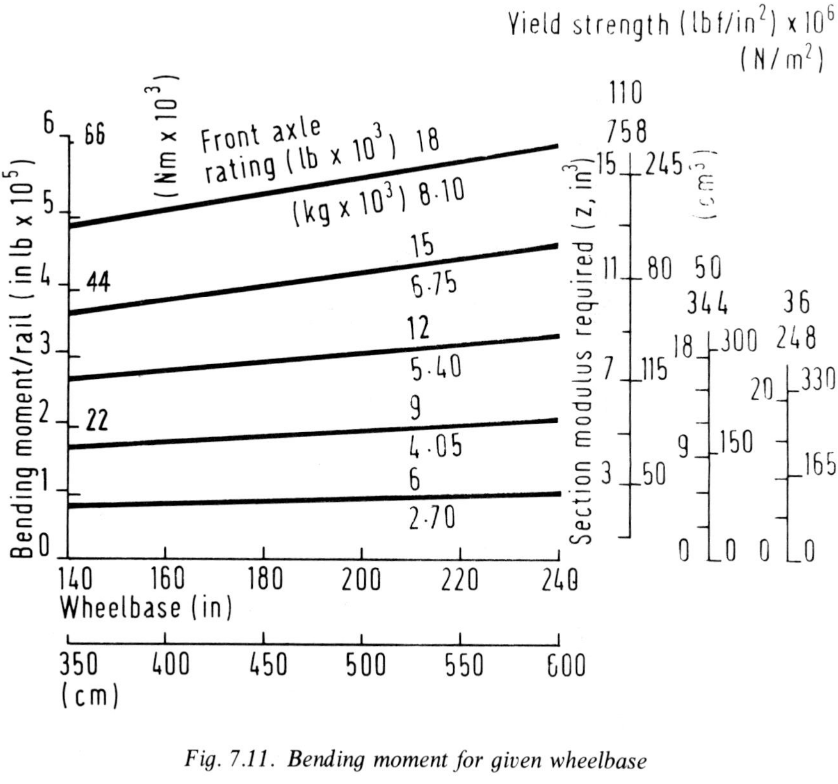

Fig. 7.11. Bending moment for given wheelbase

design criteria as 0·5 in maximum deflection between front and rear axles and allowable static stress as 0·33 times yield strength. The problem then is to translate various combinations of wheelbase, axle ratings, and load applications into side member requirements such as section modulus and second moment of area.

Figures 7.11 and 7.12 show bending moment and second moment of area required versus wheelbase for front axle ratings of 6000–18 000 lbf (2722–8165 kg). Each line on the graph represents the upper boundary of a band including all rear axle combinations for the indicated front axle rating; evenly distributed loading is assumed. With axles loaded to capacity front axle rating has more influence on frame requirements than does the rear. Right-hand side of the graph relates section modulus requirements for the three material strength levels to the maximum bending moment.

Channel section profiles are then examined in detail using a 0·25 in (6·357 mm) thick, 9 in (220 mm) deep, and 3 in (76·2 mm) flange width section as datum. Figure 7.13 shows the effect of the three variables on weight. Section properties for the datum channel are $I = 68{\cdot}07\ \text{in}^4$

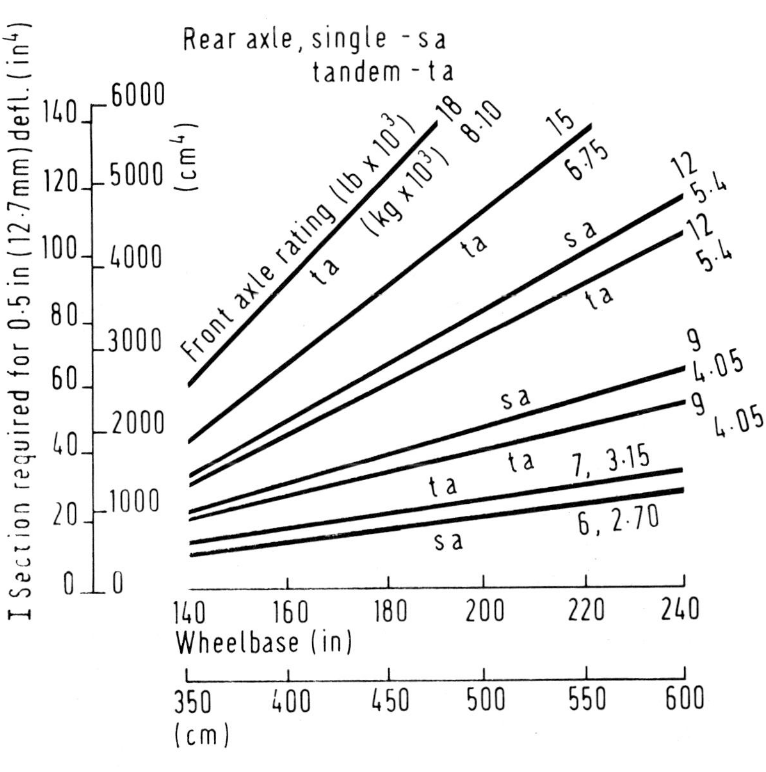

Fig. 7.12. Section second moment of area for given wheelbase

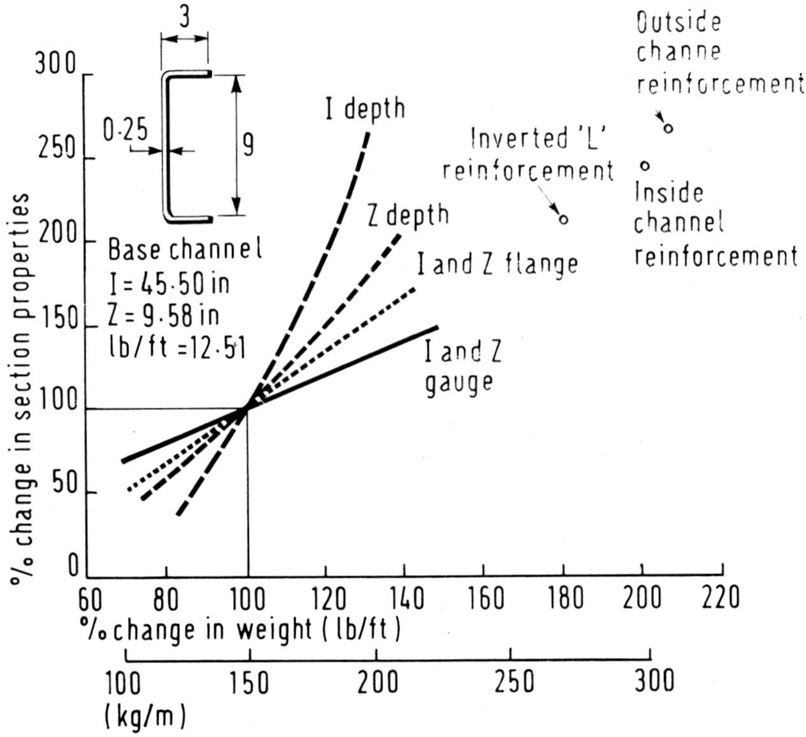

Fig. 7.13. Change in weight for various section properties

(2840 cm^4), $Z = 13{\cdot}96$ in^3 (228 cm^3), and weight per foot = 18·55 lbf (8·3 kg) for a 10 per cent increase in weight; the effect of varying the three variables is as follows:

	Weight	*Mom. inertia*	*Sec. modulus*
Depth	+10	+43	+24
Flange width	+10	+18	+18
Gauge	+10	+10	+10

From a practical standpoint the author claims that flitching is employed for section second moment of area requirements of $I < 50$ in^4 (2080 cm^4). Increase in section properties for different flitching methods is shown in Fig. 7.13, versus increase in weight.

Weight penalties in flitching prompts examination of variable depth side members. Figure 7.14 shows plots of section second moment of area against depth of the channel for different flange width and gauge. Procedure in side member selection is to make full advantage of depth permissible for a straight frame with minimum gauge. Section properties

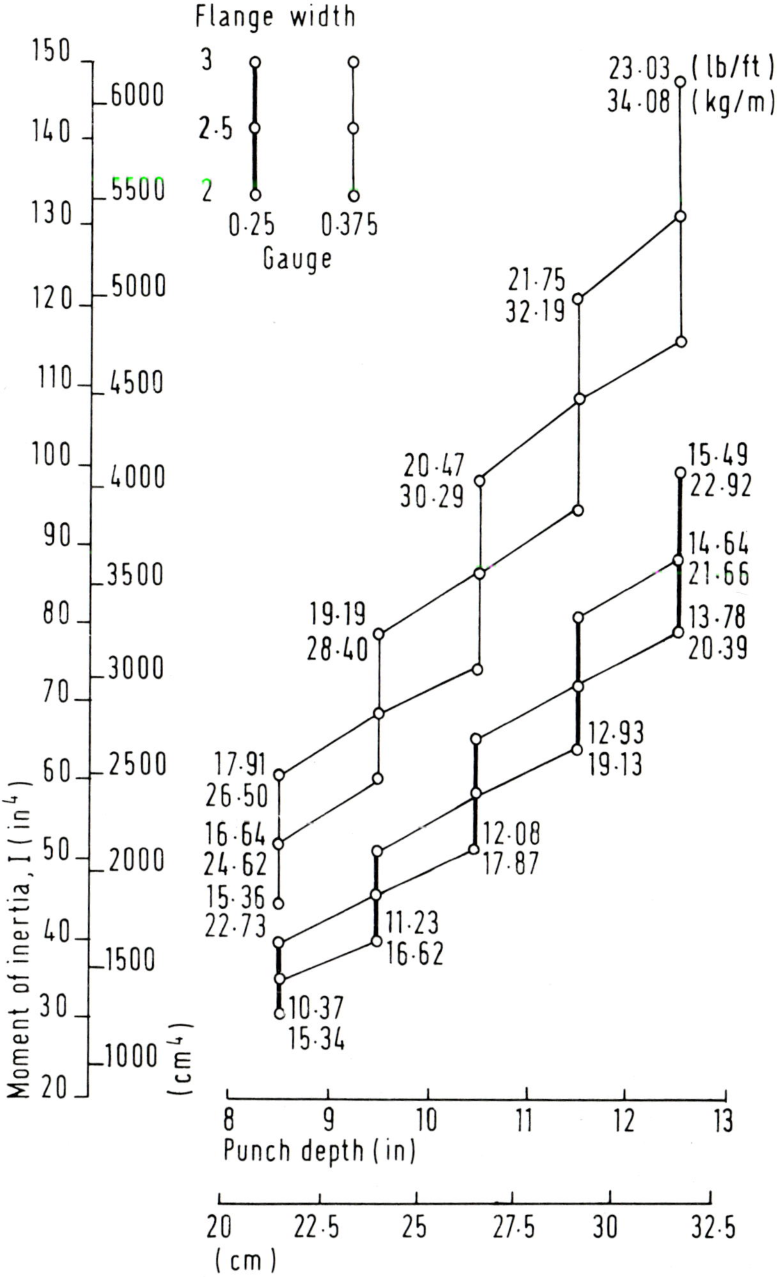

Fig. 7.14. Section second moment of area for different channel depths

are obtained from graphs like Fig. 7.14 normally based on 1/32 in gauge increments.

Reference is now made to Fig. 7.11 on which the *I* value just obtained is drawn as a horizontal line and indicates the application for which the section will equal or exceed the requirement.

LATERAL INSTABILITY OF SIDE MEMBERS

Tests carried out by Lewis McNitt reporting in SAE paper 690 176 have established the importance of designing-in frame side member stability to fully exploit chassis bending strength. Chassis frame specimens, 15 ft (4·57 m) long, typical of artic tractor configuration, were loaded through a typical fifth wheel coupler mounting and strain gauge readings were recorded at various positions around the side member channel sections at mid span. The effect of three different cross member spacings was examined and measurements of lateral deflection of the section measured as an indication of sideways buckling. Lateral deflection of the channel section flange was found to be in the order of 2 in (50 mm) for a 9·5 in (242·5 mm) deep section (with 3 in (76·2 mm) flanges and 0·25 in (6·35 mm) metal thickness) for large cross member spacing.

Figure 7.15 shows the strain gauge reading at the open edge of the

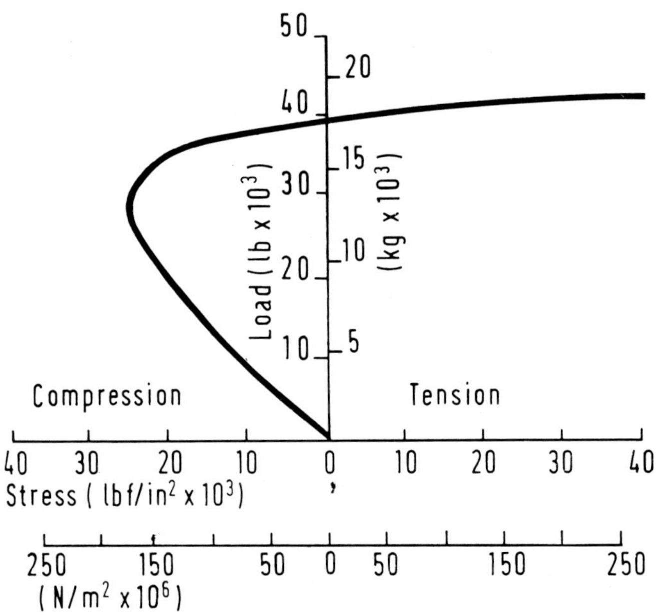

Fig. 7.15. Strain gauge readings on the open edge of a channel flange

channel flange with 96 in (2·44 m) cross member spacing. As the member is loaded a linear compressive stress/applied load is the case up to 28 000 lb (12 701 kg), after which lateral bending of the flange causes a reversal in stress at around 40 000 lb (18 144 kg) when the 'torsional' effect overcomes vertical bending and a state of tensile stress occurs.

Figure 7.16 shows the stress distribution across the section with 66 in (1·676 m) cross member spacing. Calculated stresses (using Engineers Theory of Bending) are shown dotted. At the higher loads it is clear that stress distribution across the flange is not uniform as calculated from simple theory. In fact, the radius area yields while the remaining portion of the flange continues to take higher stress until failure when the flange edge stress reaches yield. Maximum shear occurs at the radius edge and reduces in a triangular distribution to zero at the edge of the flange.

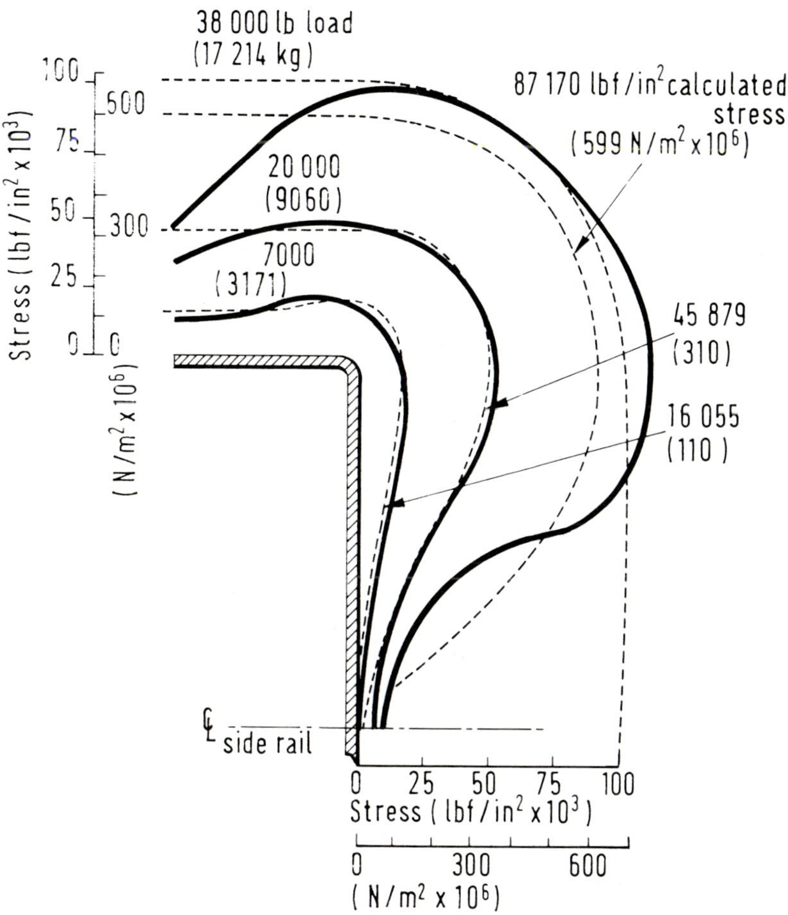

Fig. 7.16. Stress distribution across a channel section

With the shorter cross member spacing, 36 in (0·914 m) instead of 66 and 96 in (1·676 and 2·44 m), failure was due to flange buckling rather than lateral instability but 23–32 per cent greater load carrying capacity was obtained. Figure 7.17 shows the angular deflection of the side member section at A 36 in (0·914 m), B 66 in (1·676 m), and C 96 in (2·44 m).

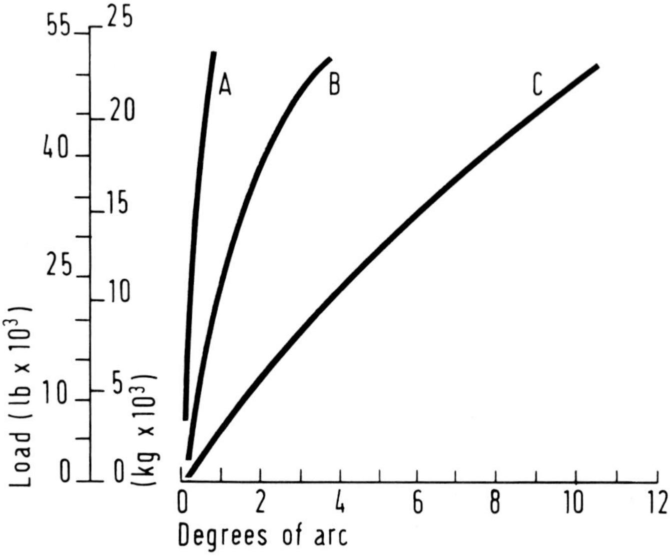

Fig. 7.17. Angular deflection of a side member section

While the above tests were made with the load applied directly on to the side member top flange, further tests were made with the load applied at the shear centre of the channel section. Figure 7.18 shows the higher loads that were attainable for shear centre application on the basis of equal stress in the radiused end of the flange.

A move to stiffer frames?

D. W. Sherman of the Dana Corporation writing in SAE paper 690 175 considers the detail design of side member attachments. The familiar alligator cross member he explains is designed to give good joint rigidity without severe stress concentration, caused by restraining axial warping, while giving overall frame resilience. The latter is obtained by exploiting the flexibility of moderate sections over the main span of the cross member.

Twist applied to chassis frames when vehicles traverse a single wheel bump is accompanied by lateral displacement of the axle which in turn, through the suspension spring, creates a severe bending moment on the spring hanger bracket. The limiting effect on the side member is shown in Fig. 7.19 for the case of a flange-connected alligator cross member coincident with the spring hanger brackets. The spring bracket imposes a web

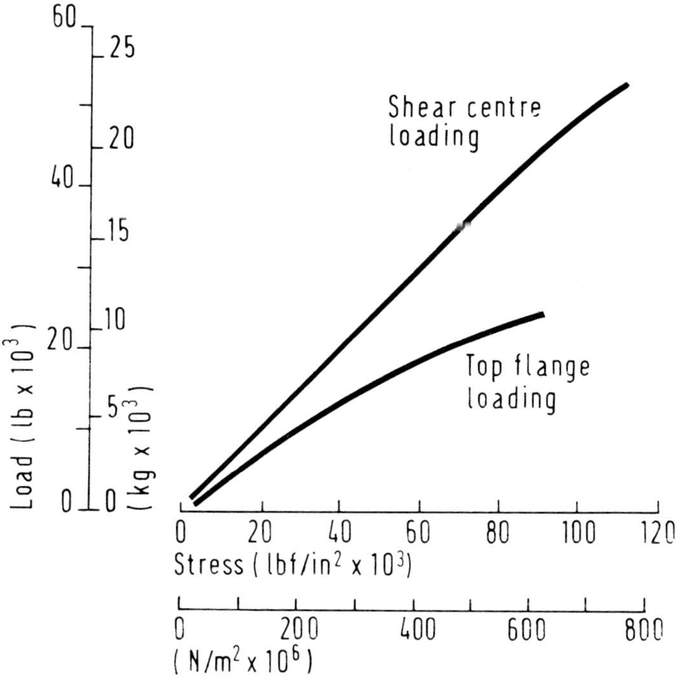

Fig. 7.18. Higher loading attainable by aligning with the shear centre

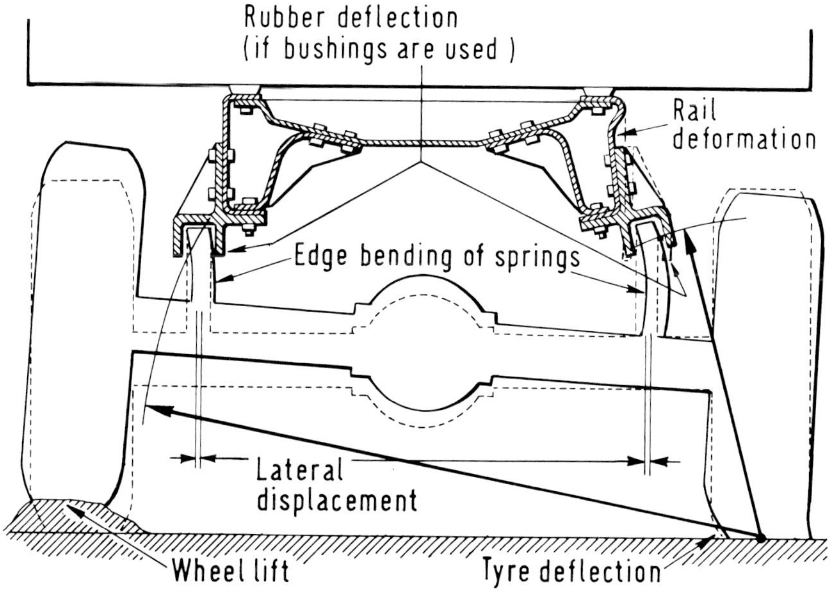

Fig. 7.19. 'Oil-canning' of chassis side member

load resulting in diaphragming of the web. This can be alleviated by extending the spring bracket upwards as far as possible.

Writing in *Technische Rundschau*, number 16/62, W. Seitler of the Swiss Company Adolph Saurer has considered the need to raise the torsional stiffness of chassis frames, providing some quantitative values and suggesting ways of reducing stress concentration. He points out that if the chassis frame is to fulfil its secondary role of maintaining wheel to ground contact then softer suspension springs have to be provided as the frame is torsionally stiffened.

For the arrangement in Fig. 7.20 the wheel deflection which will just lift the front of the vehicle is given by $h = F_1 \cdot f = W_v \cdot f \cdot l_g/l$ where f is the sum of the flexibilities of front suspension f_f, rear suspension f_r, and frame f_{ch} given by

$$f_f = \frac{1}{K_{fs}}\left(\frac{t_{fw}}{t_{fs}}\right)^2 + \frac{1}{K_{ft}}$$

$$f_r = \left(\frac{t_{fw}}{t_{rw}}\right)^2 \left\{\frac{1}{K_{rs}}\left(\frac{t_{rw}}{t_{rs}}\right)^2 + \frac{1}{K_{rt}}\right\}$$

$$f_{ch} = \frac{t_{fw}^2}{2K_{ch}}\frac{\pi}{180}$$

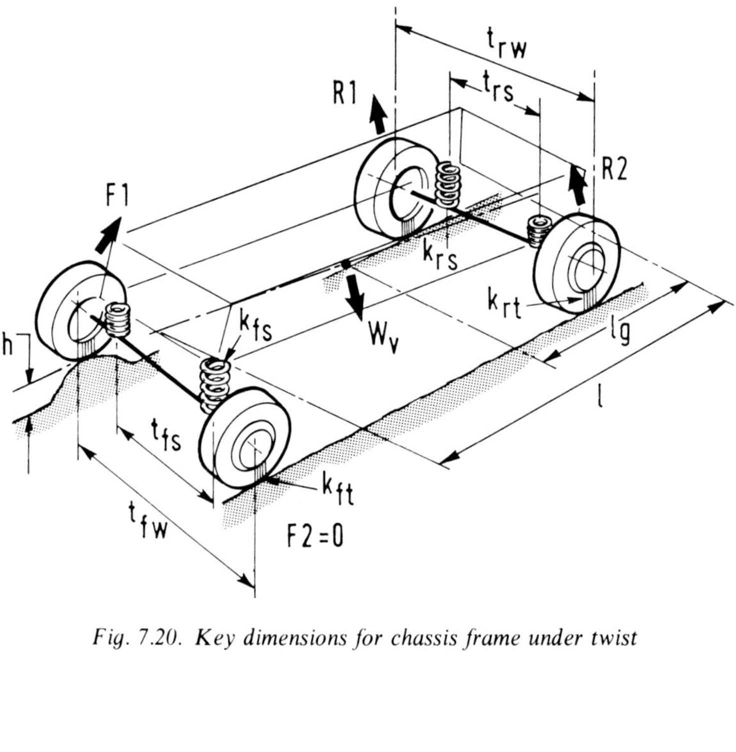

Fig. 7.20. Key dimensions for chassis frame under twist

Typical values for trucks are $f_f = 0{\cdot}0032$–$0{\cdot}00535$ in/lbf (0·01584–0·0308 mm/N), $f_r = 0{\cdot}0035$–$0{\cdot}00125$ in/lbf (0·02015–0·0072 mm/N), $f_{ch} = 0{\cdot}0024$ in/lbf (0·0138 mm/N) for a conventional frame and $f_{ch} = 0{\cdot}00057$ in/lbf (0·00327 mm/N) for a welded tubular cross member frame. For a uniform load of 8820 lb (39·2 kN) on the front axle and 197 in (5 m) wheelbase, $h = 4{\cdot}6$–$8{\cdot}5$ in (117–216 mm) for the conventional frame and 3·6–6·7 in (91·7–170 mm) for the tubular cross membered frame.

When efforts to raise frame torsional stiffness were made with open channel cross members, severe stress concentrations resulted. Tests carried out have shown that tubular cross member frames are some 2·8 times stiffer than conventional ones, when both are subject to 0·5 degree twist, and have 80 per cent lower normal stresses than the conventional frames.

For specialist earthmoving trucks such as the Aveling Barford Centaur, high wheel displacement suspensions are built in; in this case wheels are sprung by oleo-pneumatic units. Consequently high stiffness frames are used with stout circular tubed cross members of 12 in (305 mm) dia, see Fig. 7.21. Side members, too, are of box section, with integral front bumper. I-section 'peripheral' cross members encircle the power/transmission unit.

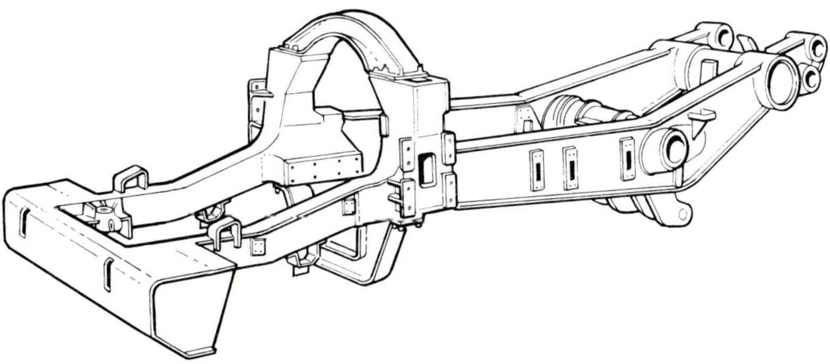

Fig. 7.21. Chassis frame of Aveling Barford Centaur

Aluminium alloy chassis frames that have been used in semi-trailer construction also feature rigid frame construction. In the Duramin coil-carrying trailer shown in Fig. 7.22 the side members comprise high-strength alloy profiled webs, maximum depth 18 in (457 mm), enclosed by extruded tee-section flanges, Avdelok riveted in position. Generous flitching is used at the tapered portions of the side members and tapered channel is used for outriggers and cross members. Basic structural weight, including deal flooring, is 22 cwt (1·12 tonne) for a 22 ft (6·7 m) trailer.

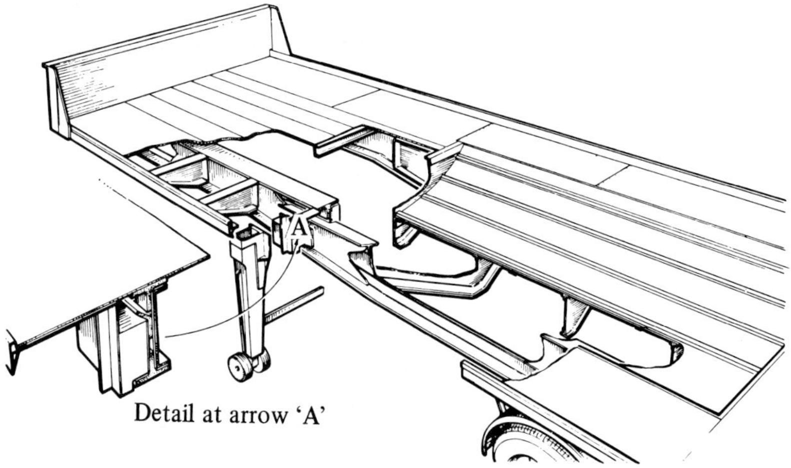

Fig. 7.22. Duramin aluminium alloy trailer chassis

In the USA three major truck builders use aluminium alloy frames. With the weight advantage of aluminium it is usually possible to be more generous at member junctions, and other areas involving severe discontinuities, where fatigue could occur. Generally a non-copper bearing aluminium alloy such as H30 is preferred in chassis-frame constructions for its resistance to corrosion.

BOX VAN CONSTRUCTION

Van body structures range from the car-derivative type, through panel van and chassis-mounted box van to semi-trailer box vans. The car-derivative van is virtually an estate car with panelled-in sides normally built on to integral structured cars. The Hillman Imp is rather a special example with rear engine location, Fig. 7.23. Due to the inclined engine the cargo space floor is not exceptionally high at $27\frac{1}{4}$ in (695 mm). There is 50 ft^3 (1·416 m^3) of volume behind the driver, 6 ft^3 (0·17 m^3) below the cargo floor, 14 ft^3 (0·396 m^3) alongside the driver and 3·25 ft^3 (0·915 m^3) in the front end 'boot'.

Of the panel-van category the Ford Transit is a good example shown in Fig. 7.24. Underbody longitudinal members are 3·9 in (99·2 mm) deep by 3·25 in (82·7 mm) wide inverted top hat sections of 0·052 in (1·33 mm) thickness spaced 37·69 in (960 mm) apart. The open top of the hat is enclosed by the floor plate. There are five main cross members again of inverted top hat section enclosed by the floor plate.

In the chassis-mounted box-van category, hand-made (coachbuilt)

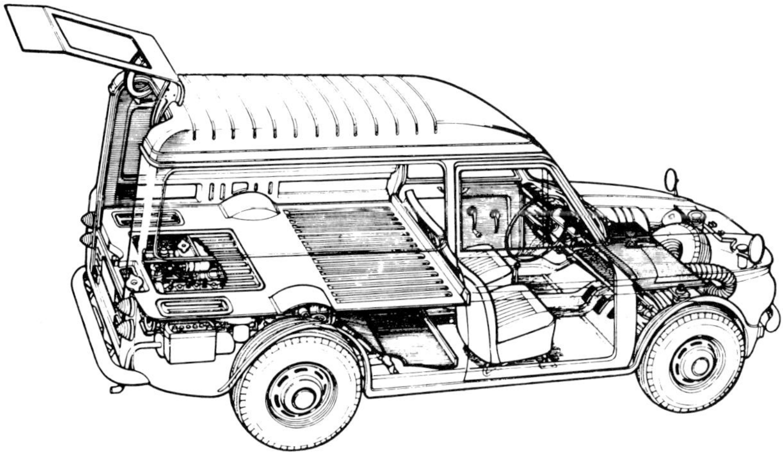

Fig. 7.23. Hillman Imp van

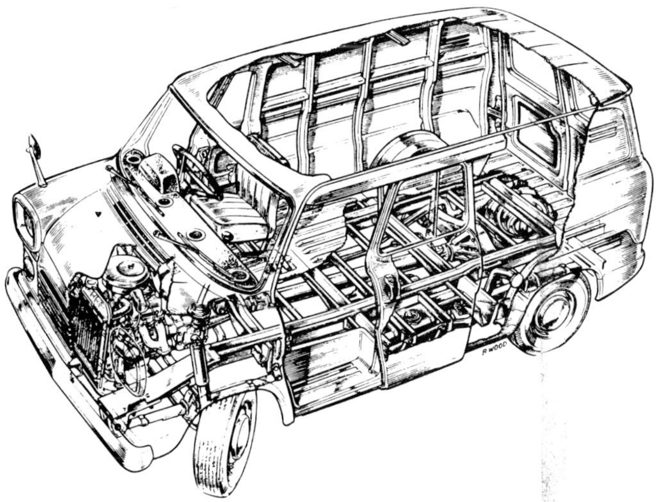

Fig. 7.24. Ford Transit van

bodies are the norm. Among the best of these is the Tidd Strongbox shown in Fig. 7.25, panelled in grp faced marine ply on an extruded aluminium frame. Corner pillars feature built-in coving; cant rails incorporate an exterior roof fixing flange to minimize leakage and an interlocking rubbing rail that covers the fasteners. The main side rail (sole bar) is supported on 102 mm deep I-section cross bearers.

The box-van semi-trailer has much in common with the cargo container and is usually built in 40 or 20 ft (12·2 or 6·1 m) lengths to meet material

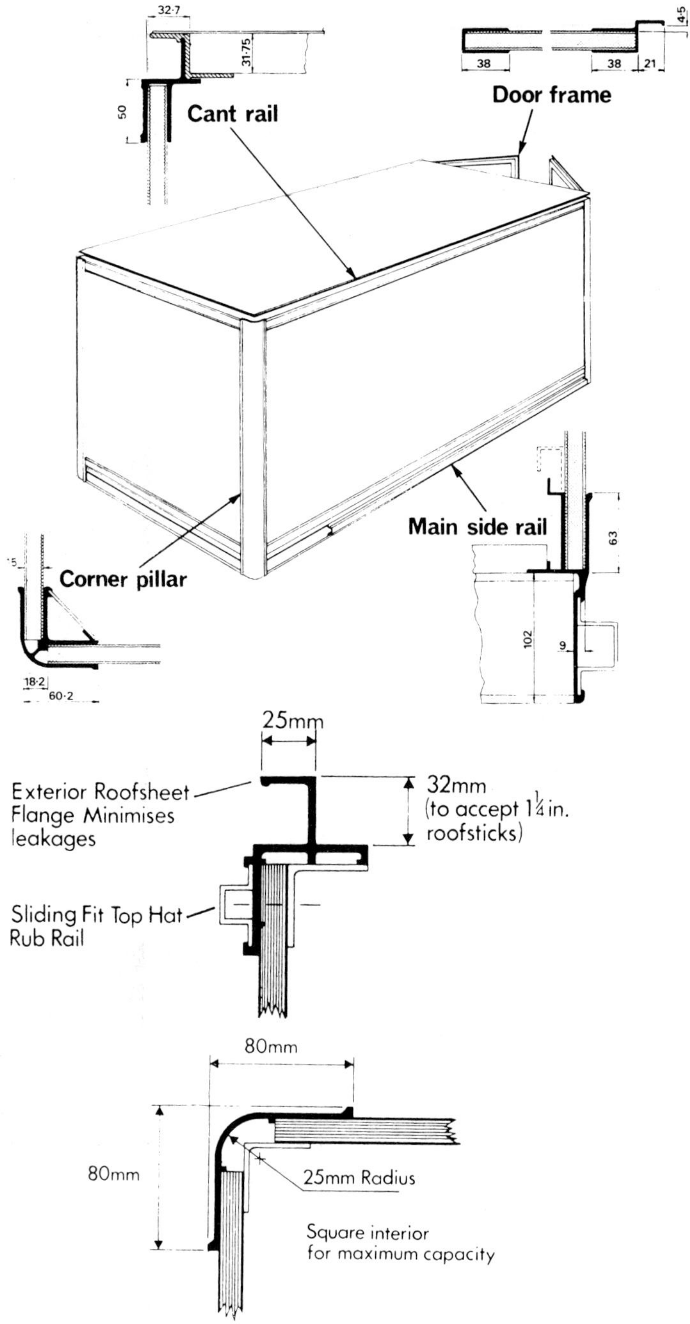

Fig. 7.25. Tidd Strongbox box van construction

handling standards. A typical floor loading criterion used is the individual wheel loading (1650 lbf or 7·33 kN) of a material handling truck. The handbook *Commercial Vehicle Bodybuilding in Aluminium* published by British Aluminium Co. recommends floor plate thickness for given distributed loading in lbf/ft^2 and crossbearer spacing in inches as in Table 7.2, first in relation to a maximum deflection criterion and second to a working stress. Floor plate thickness given in the body of the table is in swg and fractions of an inch.

TABLE 7.2

Floor thickness (swg) for given loading (lb/ft^2) uniformly distributed

Crossbearer spacing (in)	50	75	100	150	200	250
15	10	10	10	10	9	8
18	12	10	9	8	$\frac{3}{16}$ in	$\frac{3}{16}$ in
21	10	9	8	$\frac{3}{16}$ in	$\frac{7}{32}$ in	$\frac{7}{32}$ in
24	9	8	$\frac{3}{16}$ in	$\frac{7}{32}$ in	$\frac{1}{4}$ in	$\frac{1}{4}$ in
27	8	$\frac{3}{16}$ in	$\frac{7}{32}$ in	$\frac{1}{4}$ in	$\frac{9}{32}$ in	$\frac{9}{32}$ in
15	10	10	10	10	10	8
18	10	10	10	8	8	$\frac{3}{16}$ in
21	10	10	10	8	$\frac{3}{16}$ in	$\frac{7}{32}$ in
24	10	10	8	$\frac{3}{16}$ in	$\frac{7}{32}$ in	$\frac{7}{32}$ in
27	10	8	8	$\frac{7}{32}$ in	$\frac{1}{4}$ in	$\frac{1}{4}$ in

An advanced design proposal for an insulated semi-trailer container body is shown in Fig. 7.26; it is a sandwich construction based on an expanded PVC core material of 'structural density'. Here the floor construction is aluminium alloy tread plate over a plywood load spreader on to the expanded PVC core. The bottom surface of the floor sandwich panel rests on crossbearers.

Suspension of items from the sandwich panel roof is by $1\frac{1}{2}$ in (38 mm) dia, 3 in (76 mm) long Tufnol plugs having ultimate tensile strength of 500 lbf (2·2 kN) and 100 lbf (445 N) working strength. The wall panels require no supporting pillars and the aluminium alloy skins can be worked to 14 000 lbf/in^2 (96·5 MN/m^2) before local instability of the recommended expanded PVC core material occurs. Even allowing for a safety factor of five a laminate with two 20 swg skins can support an edgewise load of 1 ton/ft (3333 kg/m).

The design and analysis of chassisless semi-trailer vans has been well described by Leonard Bouge writing in SAE paper 660 077 who describes the performance of a van structure as follows. The floor system (planks and

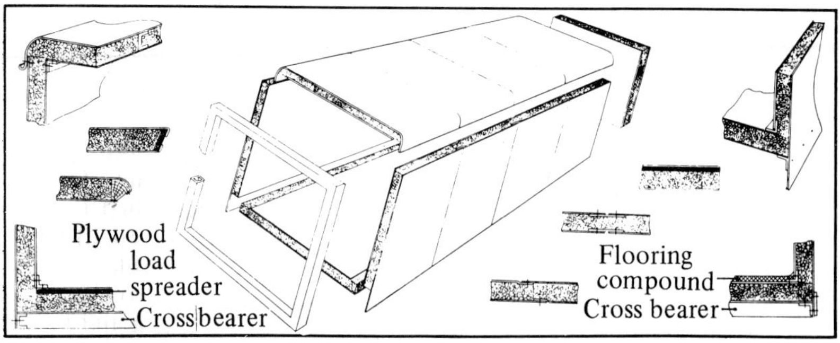

Fig. 7.26. Semi-trailer van body proposal for Plasticell core material

cross members) transmits the load to the side panel. The side panel acts as a beam, transmitting the load to the coupling gear at one end and the suspension at the other. Twisting is resisted by the entire body acting as a tube and primarily by the end frames. Horizontal forces are first resisted by the understructure. Bending in the horizontal plane, occurring on cornering, is resisted by the floor system and roof assembly. Figure 7.27 provides a nomenclature for the structure.

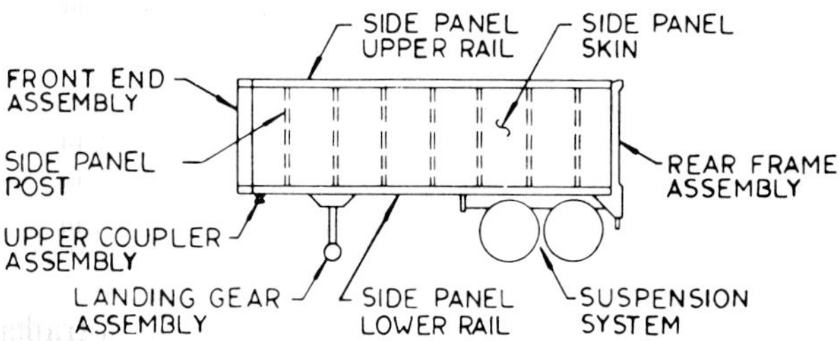

Fig. 7.27. Semi-trailer van structural nomenclature

The upper rail is a compression member, similar to the upper flange of an I-beam, in addition to supporting the roof bows. The lower rail is a tension member, similar to the lower flange of an I-beam, in addition to supporting floor crossbearers. The posts are compression members which keep lower and upper rails apart and limit skin buckling, as well as providing a fastening surface for the liner. The primary structural function of the skin is to stabilize (or maintain the squareness) of the frame structure. It is a load carrying member in which a limited degree of buckling is allowable. It is analysed structurally by semi-tension field design.

Sandwich panel analysis

Sidepanel strength in van bodies is influenced primarily by the resistance to buckling. The most efficient panel construction, therefore, is one in which the buckling strength reaches the compressive yield stress of the skin material. A typical coachbuilt panel with pillars spaced at 2 ft (610 mm) pitch and horizontal rails at say, 3 ft (916 mm) pitch would buckle at only 15 per cent of the compressive yield stress of the panel. Even highly sophisticated sheet/stringer designs used in aircraft production would be unlikely to be more than 50 per cent efficient on the same basis. A sandwich panel, however, of appropriate face and core material properties and dimensions could theoretically be 100 per cent efficient.

One of the prime advantages of the sandwich panel is its high specific section second moment of area I, brought about by the position of the load-bearing face skins at the extremities of the section, separated by a low density stabilizing core. It is important though to obtain panel dimensions and shear strength in the core that fully exploit the strength of the face skins.

Consider first instability under end loading, Fig. 7.28; this condition can occur both in overall column-buckling and in local 'wrinkling' of the faces. Local wrinkling has been considered by Dai Williams writing in his 'Structures Report No. 2' of the Royal Aircraft Establishment. He used a strain energy method to evaluate end load corresponding to the onset of wrinkling in the faces, which were considered to be elastically supported, the

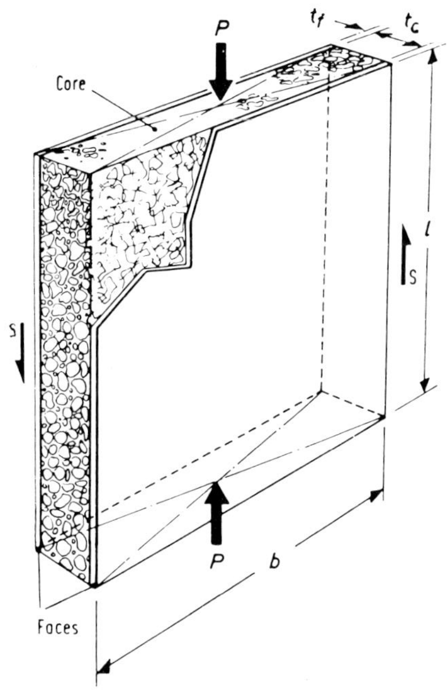

Fig. 7.28. Conventional notation for sandwich panels

support of one being influenced by the presence of the other. He obtained the critical wrinkling stress: $\sigma_{cr.w} = 0{\cdot}63E_f^{1/3} \,.\, E_c^{2/3}$ where E_f and E_c are elastic moduli of face and core, respectively. For overall column buckling, bending rigidity ($E_f I$) is the important parameter which is also affected by the core flexibility. The problem can be analysed in a way similar to that used for the consideration of latticed struts.

Writing in *Theory of Elastic Stability* (McGraw-Hill Books), Timoshenko and Gere say that the critical buckling load P_{cr} for an unsupported 'column' having comparatively low shear stiffness should be factored by $1/(1 + n \,.\, P_{cr}/A \,.\, G)$ where A is the cross-section area, G is shear modulus, and n is ratio of maximum to average shear stress, which depends on the shear distribution across the section. For a sandwich panel, if parabolic shear distribution across the section is taken to exist then $n = 3/2$. So the crippling load due to shear stiffness in bending alone can be written $P_{cr.s} = A \,.\, G/n$ and due to bending stiffness alone

$$P_{cr.b} = \pi^2 \,.\, E_f \,.\, I/1^2(1 - v_f^2)$$

where $1/P_{cr} = 1/P_{cr.b} + 1/P_{cr.s}$. Poisson's ratio for the skin material $= v_f$. The section second moment of area of the sandwich panel can be simplified, by neglecting the face skin inertias about their own axes as $I = t_f \,.\, t_c^2/2$ for unit breadth of panel and t_f and t_c thicknesses of face and core.

Stability in shear should next be considered. Again Timoshenko provides an expression for critical shear flow for a rectangular plate as $q_{cr} = K_s \,.\, \pi^2 \,.\, E \,.\, I/b^2$ for panel width b and K_s taking values from Fig. 7.29. For flexible cored sandwich panels, q_r should be factored by $1/(1 + E_f I/t_c C_s)$ where $C_s = K_{sp} b^2 \,.\, G_c/\pi^2$, where G_c is the core shear modulus and K_{sp} is found in Fig. 7.29. These last expressions combine to give critical shear stress

$$\tau_{cr} = \frac{2{\cdot}75 \,.\, K_s \,.\, E_f \,.\, t_c^2/b^2}{1 + 5{\cdot}5E_f t_f t_c/(b^2 \,.\, K_{sp} \,.\, G_c)}$$

As a worked example, a steel-faced material is considered with working stress of 40 000 lbf/in^2 (276 MN/m^2) and $E_f = 30{\cdot}10^6$ lbf/in^2 (206 GN/m^2) so that substitution in the wrinkling stress equation gives required core elastic modulus $E_c = 3000$ lbf/in^2 (20·6 MN/m^2) for 100 per cent stability up to the working stress.

A typical roof panel in compression is considered, measuring 12·5 × 7·5 ft (3·815 × 12·285 m) with a face thickness of $t_f = 0{\cdot}03$ in (0·76 mm) chosen to give resistance to local accidental damage. Then for overall column buckling, the crippling load would be factored by 14·3 for assumed

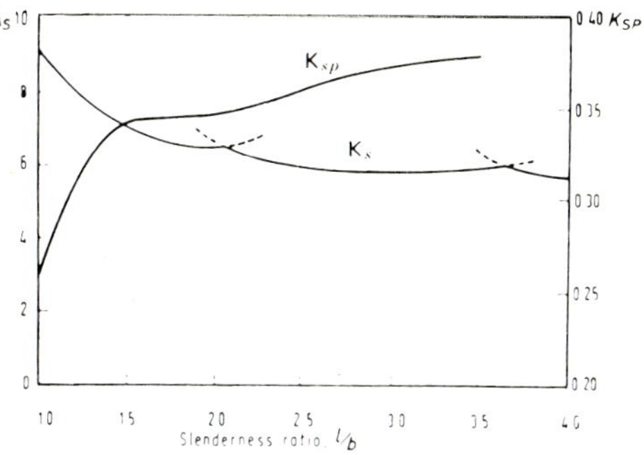

Fig. 7.29. Buckling stress coefficients for sandwich panels

simple support of the roof by the side panels. Substituting in the column buckling expression with G_c taken as 1150 lbf/in^2 or 7·93 MN/m^2 (corresponding to $E_c = 3000$ lbf/in^2 or 20·6 MN/m^2)

$$\frac{14{\cdot}3}{\sigma_{cr\,.\,b} \times 2 \times 0{\cdot}03} = \frac{1{\cdot}82 \times 12{\cdot}5^2 \times 144}{10 \times 30 \times 10^6 \times 0{\cdot}03 \times t_c^2} + \frac{3}{2 \times 1150 \times t_c}$$

which reduces to

$$\sigma_{cr\,.\,b} = \frac{238\,000}{(4{\cdot}54/t_c^2) + (1{\cdot}305/t_c)}$$

This can be plotted on a panel stability graph as in Fig. 7.30.

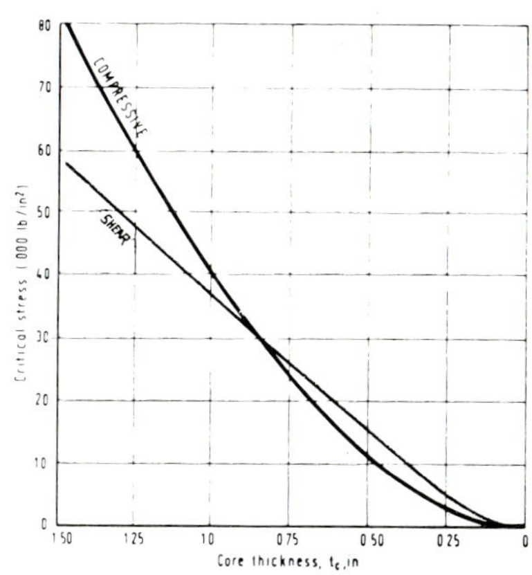

Fig. 7.30. Stability curves for a sandwich panel

Finally, we consider van sidewall panels measuring 12 × 6 ft (3·66 × 1·83 m), with central concentrated load reacted by end frames, giving effectively two 6 ft (1·83 m) square panels under action of shear forces. So that substitution in the shear stability formula

$$\tau_{cr} = \frac{2{\cdot}75 \times 9{\cdot}3 \times 30 \times 10^6 \times t_c^2/72^2}{1 + (5{\cdot}5 \times 30 \times 10^6 \times 0{\cdot}03 \times t_c)/(72^2 \times 0{\cdot}255 \times 1150)}$$

since $K_s = 9{\cdot}3$ and $K_{sp} = 0{\cdot}255$ from Fig. 7.29. Therefore $\tau_{cr} = (14{\cdot}8 \times 10^4 \times \tau_c^2)/(1 + 3{\cdot}25t_c)$ which again is plotted in Fig. 7.30. These curves can then be used to determine core thickness for desired operating stress levels.

Box beam behaviour

In assuming optimum structural efficiency for the box shell structure of the van, the behaviour of box shell beams needs to be examined. We know from Chapter 3 that the conditions for which the Engineers Bending Theory can apply are that plane sections remain plane (no axial warping), that cross sections retain their planar shape after bending (no swelling or bowing-in), and that load is applied uniformly over the section.

But for the loaded cantilever box beam in Fig. 7.31 we see that at least one of the above conditions is violated. Because of the parabolic distribution of shear stress across the vertical webs of the box beam, 'essing' of the

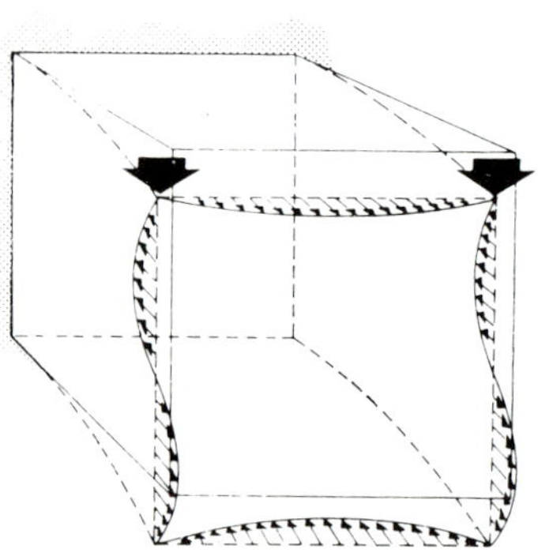

Fig. 7.31. Sectional warping of a box beam

sidewalls takes place. There is also axial warping of the horizontal 'flanges' caused by shear lag. This is because panel flexibility in shear means that bending forces applied at the panel edges are prevented from retaining an even value across the panel width.

Practical box beams are usually reinforced by various forms of stiffening to stabilize the thin walls. Nomenclature used in shell theory for the various members is shown in Fig. 7.32. Dai Williams in *The Theory of Aircraft Structures* (Arnold) shows that stiffeners can seriously affect the valid-

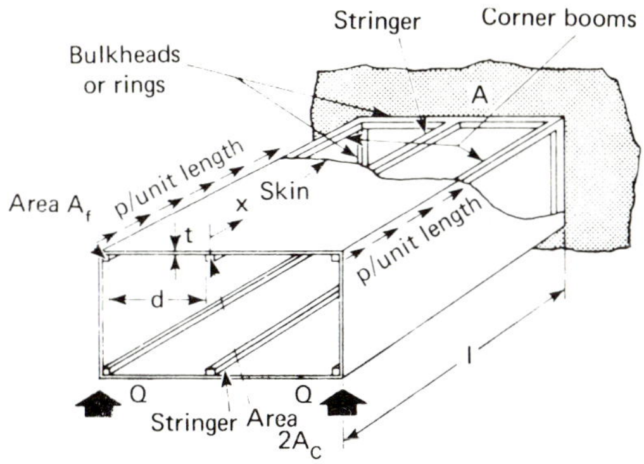

Fig. 7.32. Nomenclature for reinforced box beams

ity with which the Engineers' Theory of Bending can be applied to a box beam. With the assumption that the skin sheet can withstand shear only, and that the roof panel is under shear flow only from vertical loading Q, he shows that stringer and boom loads at x from the free end are given by

$$\text{Boom load} = \left(\frac{P}{A_f + A_c}\right)\left(A_f \,.\, x + \frac{A_c}{k}\frac{\sinh \,.\, kx}{\cosh kl}\right)$$

$$\text{Stringer load} = \left(\frac{P}{A_f + A_c}\right)\left(A_c \,.\, x - \frac{A_c}{k}\frac{\sinh \,.\, kx}{\cosh kl}\right)$$

$$\text{Shear flow in skin} = \left(\frac{P \,.\, A_c}{A_f + A_c}\right)\left(1 - \frac{\cosh kx}{\cosh kl}\right)$$

where

$$k = \sqrt{\left(\frac{Gt}{Ed}\frac{A_f + A_c}{A_f A_c}\right)}$$

and E and G are elastic and shear moduli for the beam material. The second, hyperbolic terms in these expressions give the modification of the loads obtained by simple beam theory due to shear lag, there being an increase and decrease in loading for the boom and stringer, respectively.

Putting the following numerical values for the stiffener parameters: $d = 20$ in (508 mm), $l = 100$ in (2·54 m), $A_f = A_c = 4$ in (25·8 cm^2), $t = 0{\cdot}04$ in (10·17 mm) and $G/E = 0{\cdot}4$, Williams obtained the results shown in Fig. 7.33. This indicates the order of departure of shear lag theory (full lines) from Engineers theory (broken lines). For a long beam, however, since $\cosh kx/\cosh kl \to 0$ as $l/\cosh kl \to \infty$, Engineers' Theory becomes progressively more accurate.

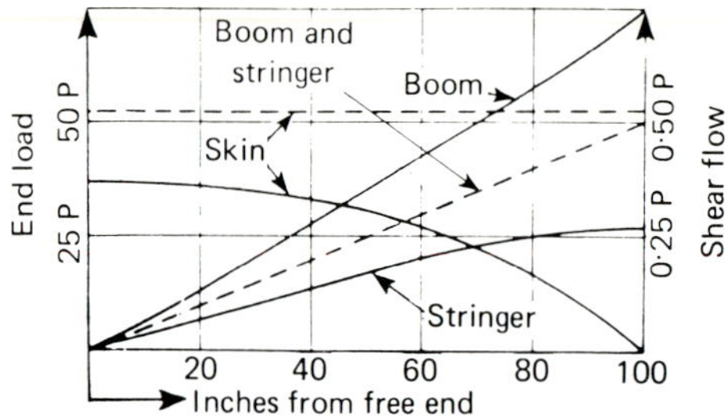

Fig. 7.33. Effect of shear lag on box beams

APPLICATION OF MATRIX METHODS

The unit load method has been established in an earlier chapter in terms of flexural deformation, after which an example was given of applying the method to determine deflections of a structure due to flexural, torsional, shear, and generalized end load. In this last case, deflection of a determinate structure is given by $\delta = \sum S_0 \,.\, b_1 \,.\, l/AE$ where S_0 is member force due to real external load, b_1 is member force due to imaginary unit load in direction of δ to be determined, and l/AE is member flexibility, f. (l is the member length, E is the elastic modulus, A is the cross-section area.)

For redundant structures an extra equation is required for each degree of redundancy, to obtain a solution. This is obtained from the condition of compatibility of strain, whereby the structure has its redundant member cut and another unit load applied in the direction of the force in the cut member; the deflection at the cut is determined by the unit load method and equated to zero or the initial condition of strain. The member forces in the actual redundant frame are given by $S = S_0 + X \,.\, b_1 = b_1 \,.\, R + b_1 \,.\, X$ for externally applied load R. The condition for zero deflection at the cut is given by

$$\sum b_1^2 \,.\, f \,.\, X = -\sum b_0 \,.\, R \,.\, b_1 \,.\, f$$

The transformation of the Unit Load method of analysis from the elementary tabular approach to one using matrices is described in an article by Alan H. Lucas entitled 'Structural analysis using matrix algebra' in *Automotive Design Engineering*, Feb. 1963, pp. 54–56. For systems of multiple redundancy the number of equations involved can be written in matrix form

$$[b_1]'[f][b_1][X] = -[b_1]'[f][b_0][R]$$

The column matrix $[b_1]$ is transposed to a row matrix $[b_1]'$ so it can be multiplied by a column matrix of flexibility $[f]$.

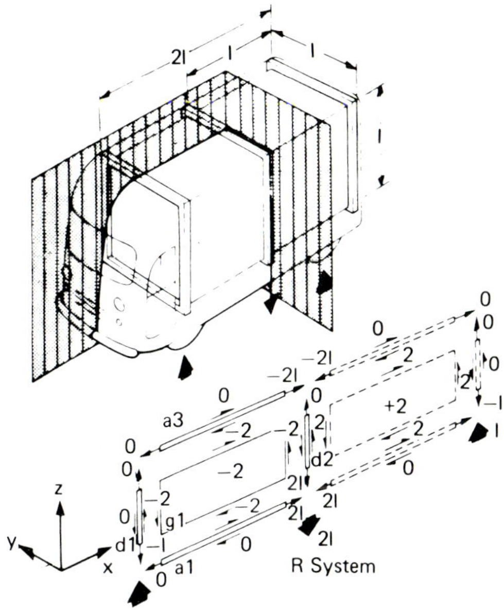

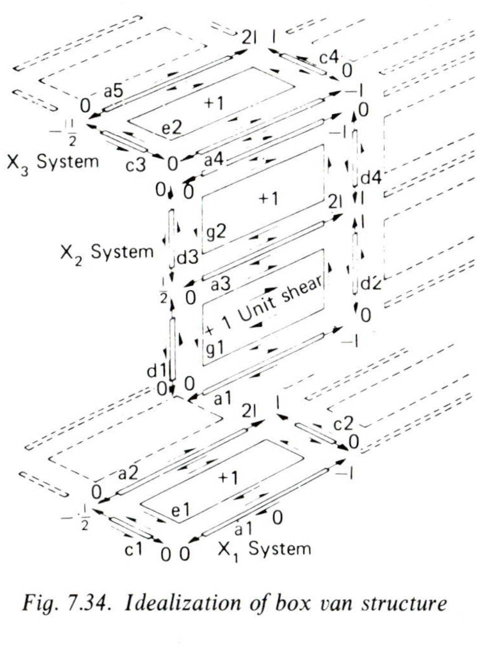

Fig. 7.34. Idealization of box van structure

And so the member forces are given by

$$[S] = [[b_0] - [b_1][[b_1]'[f][b_1]]^{-1}[[b_1]'[f][b_0]]][R]$$

Application of this method to an idealized box-van structure is shown at Fig. 7.34 where the determinate and redundant systems of plates and bars are shown. In this example concentrated loads are considered to act at the centres of the lower edges of the sidewall panels and reacted at the ends of these edges. Basically the structure comprises three bulkhead frames having pin-jointed members connected by eight sandwich panels (obviating the need for stiffeners) which are all hinge jointed. Only this simple bending case is considered, and from symmetry just one quarter of the structure needs to be examined. This quarter is further idealized into a system of simply supported plates and bars, Fig. 7.34, the bars taking the end load and the plates only the shear. These are labelled, for regularity in drawing up the matrices, according to the co-ordinate axes.

The determinate R system conveniently comprises the plate and four bars making up the lower half of the sidewalls. The three redundant systems are shown at X_1, X_2, and X_3 in the figure. The b_0 matrix is multiplied by $l/2l$, since for simplicity in algebraic expression of the matrix elements, the unit load to the R system is multiplied by $2l$. The member

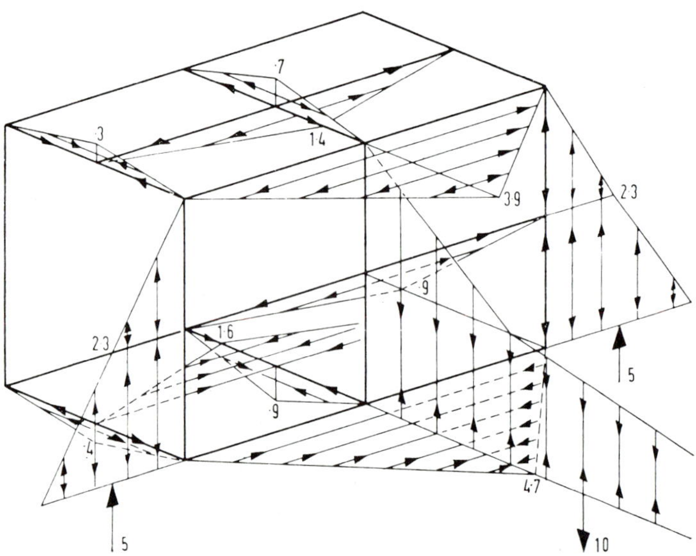

Fig. 7.36. Load distribution in a 2-bay box van body

forces are quickly determined by simple statics. In the flexibility matrix the submatrices for the bars and plates can be respectively expressed as

$$\begin{bmatrix} \dfrac{1}{3A_bE} & \dfrac{1}{6A_bE} \\ \dfrac{1}{6A_bE} & \dfrac{1}{3A_bE} \end{bmatrix} \quad \text{and} \quad \begin{bmatrix} \dfrac{A_p}{Gt} \end{bmatrix}$$

where A_b is the section area of bar, A_p is the section area of plate, t is the plate thickness and G/E is the shear/elastic moduli. In this example G is taken as $E/2$ and half the section area of the sandwich panel on each side of the equivalent 'a' bar is considered concentrated at that bar for taking end load in the longitudinal x direction. In the y and z directions the 'c' and 'd' bars, which are coincident with the bulkheads, have different section areas depending on their position either at the middle or ends of the body shell. The section area of the bulkheads is taken to be one quarter of the section area of the adjacent panel in this example. Consequently, the equivalent transverse bars at the ends take half the adjacent panel section areas, and the central transverse bars take two halves. Submatrices then become

$$f_a = \begin{bmatrix} \dfrac{2l}{3tlE} & \dfrac{2l}{6tlE} \\ \dfrac{2l}{6tlE} & \dfrac{2l}{3tlE} \end{bmatrix} \quad \text{and} \quad f_{c_1} = \begin{bmatrix} \dfrac{l/2}{3{\cdot}3tlE/4} & \dfrac{l/2}{6{\cdot}3tlE/4} \\ \dfrac{l/2}{6{\cdot}3tlE/4} & \dfrac{l/2}{3{\cdot}3tlE/4} \end{bmatrix}$$

$$f_{d_2} = \begin{bmatrix} \dfrac{l/2}{3{\cdot}5tlE/4} & \dfrac{l/2}{6{\cdot}5tlE/4} \\ \dfrac{l/2}{6{\cdot}5tlE/4} & \dfrac{l/2}{3{\cdot}5tlE/4} \end{bmatrix} \quad \text{and} \quad g = \begin{bmatrix} \dfrac{l\,.\,l/2}{t\,.\,E/2} \end{bmatrix}$$

where t is twice face thickness of the sandwich panel. The formation and manipulation of the matrices are both indicated in Fig. 7.37, the final column matrix of member forces resulting from taking an arbitary value of the external load at the bottom middle of each sidewall, $R = 10$ ton (10·16 tonne). These forces are plotted on a distribution diagram as Fig. 7.36.

The force distribution diagram shows that the method has overcome the limitations of the Engineers' Theory of bending. The effect of applying the load at the bottom of the structure instead of uniform distribution across the cross section has caused unequal compressive and tensile loads in the top and bottom rails. Reduced load in the central longitudinal members of the roof and floor, as compared with that in the side rails clearly indicates the action of shear lag.

Members		$[b_0] = \frac{1}{2l} \times$	$[b_1] =$		
$a1$	1	0	0	0	0
	2	$2l$	$-l$	$-l$	0
$a2$	1	0	0	0	0
	2	0	$2l$	0	0
$a3$	1	0	0	0	0
	2	$-2l$	0	$2l$	0
$a4$	1	0	0	0	0
	2	0	0	$-l$	$-l$
$a5$	1	0	0	0	0
	2	0	0	0	$2l$
$c1$	1	0	0	0	0
	2	0	$-\frac{l}{2}$	0	0
$c2$	1	0	0	0	0
	2	0	l	0	0
$c3$	1	0	0	0	0
	2	0	0	0	$-\frac{l}{2}$
$c4$	1	0	0	0	0
	2	0	0	0	l
$d1$	1	$-l$	0	0	0
	2	0	0	$-\frac{l}{2}$	0
$d2$	1	$2l$	0	0	0
	2	0	0	l	0
$d3$	1	0	0	$-\frac{l}{2}$	0
	2	0	0	0	0
$d4$	1	0	0	l	0
	2	0	0	0	0
$e1$		0	1	0	0
$e2$		0	0	0	1
$g1$		-2	0	1	0
$g2$		0	0	-1	0

$$[f] = \frac{1}{tE} \times \mathrm{diag}\left(\begin{bmatrix} \frac{2}{3} & \frac{1}{3} \\ \frac{1}{3} & \frac{2}{3} \end{bmatrix}, \begin{bmatrix} \frac{2}{3} & \frac{1}{3} \\ \frac{1}{3} & \frac{2}{3} \end{bmatrix}, \begin{bmatrix} \frac{2}{3} & \frac{1}{3} \\ \frac{1}{3} & \frac{2}{3} \end{bmatrix}, \begin{bmatrix} \frac{2}{3} & \frac{1}{3} \\ \frac{1}{3} & \frac{2}{3} \end{bmatrix}, \begin{bmatrix} \frac{2}{3} & \frac{1}{3} \\ \frac{1}{3} & \frac{2}{3} \end{bmatrix}, \begin{bmatrix} \frac{2}{9} & \frac{1}{9} \\ \frac{1}{9} & \frac{2}{9} \end{bmatrix}, \begin{bmatrix} \frac{2}{15} & \frac{1}{15} \\ \frac{1}{15} & \frac{2}{15} \end{bmatrix}, \begin{bmatrix} \frac{2}{9} & \frac{1}{9} \\ \frac{1}{9} & \frac{2}{9} \end{bmatrix}, \begin{bmatrix} \frac{2}{15} & \frac{1}{15} \\ \frac{1}{15} & \frac{2}{15} \end{bmatrix}, \begin{bmatrix} \frac{2}{9} & \frac{1}{9} \\ \frac{1}{9} & \frac{2}{9} \end{bmatrix}, \begin{bmatrix} \frac{2}{15} & \frac{1}{15} \\ \frac{1}{15} & \frac{2}{15} \end{bmatrix}, \begin{bmatrix} \frac{2}{9} & \frac{1}{9} \\ \frac{1}{9} & \frac{2}{9} \end{bmatrix}, \begin{bmatrix} \frac{2}{15} & \frac{1}{15} \\ \frac{1}{15} & \frac{2}{15} \end{bmatrix}, l^2, l^2, l^2, l^2 \right)$$

All zeros

One diagonal

All zeros

$$[b_1'] = \begin{bmatrix} 0 & -l & 0 & 2l & 0 & 0 & 0 & 0 & 0 & 0 & 0 & -\frac{l}{2} & 0 & l & 0 & 0 & 0 & 0 & 0 & 0 & 0 & 0 & 0 & 0 & 0 & 0 & 1 & 0 & 0 & 0 \\ 0 & -l & 0 & 0 & 0 & 2l & 0 & -l & 0 & 0 & 0 & 0 & 0 & 0 & 0 & 0 & 0 & 0 & 0 & -\frac{l}{2} & 0 & l & -\frac{l}{2} & 0 & l & 0 & 0 & 0 & 1 & -1 \\ 0 & 0 & 0 & 0 & 0 & 0 & 0 & -l & 0 & 2l & 0 & 0 & 0 & 0 & 0 & -\frac{l}{2} & 0 & l & 0 & 0 & 0 & 0 & 0 & 0 & 0 & 0 & 0 & 1 & 0 & 0 \end{bmatrix}$$

$$[b_1'][f] = \frac{1}{tE} \times \begin{bmatrix} -\frac{l}{3} & -\frac{2l}{3} & \frac{2l}{3} & \frac{4l}{3} & 0 & 0 & 0 & 0 & 0 & 0 & \frac{l}{18} & \frac{l}{9} & \frac{l}{15} & \frac{2l}{15} & 0 & 0 & 0 & 0 & 0 & 0 & 0 & 0 & 0 & 0 & 0 & 0 & l^2 & 0 & 0 & 0 \\ -\frac{l}{3} & -\frac{2l}{3} & 0 & 0 & \frac{2l}{3} & \frac{4l}{3} & -\frac{l}{3} & -\frac{2l}{3} & 0 & 0 & 0 & 0 & 0 & 0 & 0 & 0 & 0 & 0 & -\frac{l}{18} & -\frac{l}{9} & \frac{l}{15} & \frac{2l}{15} & -\frac{l}{9} & -\frac{l}{18} & \frac{2l}{15} & \frac{l}{15} & 0 & 0 & l^2 & -l^2 \\ 0 & 0 & 0 & 0 & 0 & 0 & -\frac{l}{3} & -\frac{2l}{3} & \frac{2l}{3} & \frac{4l}{3} & 0 & 0 & 0 & 0 & \frac{l}{18} & \frac{l}{9} & \frac{l}{15} & \frac{2l}{15} & 0 & 0 & 0 & 0 & 0 & 0 & 0 & 0 & 0 & l^2 & 0 & 0 \end{bmatrix}$$

$$\text{Then } [b_1'][f][b_1] = \frac{l^2}{tE} \times \begin{bmatrix} \frac{407}{90} & \frac{2}{3} & 0 \\ \frac{2}{3} & \frac{574}{90} & \frac{2}{3} \\ 0 & \frac{2}{3} & \frac{407}{90} \end{bmatrix} \qquad \text{and } [b_1'][f][b_0] = \frac{l}{2tE} \times \begin{bmatrix} -\frac{4}{3} \\ -\frac{523}{90} \\ 0 \end{bmatrix}$$

Inversion of $\left[[b_1'][f][b_1]\right]$

Factor

$$\begin{array}{c} \textcircled{1} \\ -0.15 \\ 0 \end{array} \begin{bmatrix} 4.52 & 0.667 & 0 \\ 0.667 & 6.38 & 0.667 \\ 0 & 0.667 & 4.52 \end{bmatrix} \begin{bmatrix} 1 & 0 & 0 \\ 0 & 1 & 0 \\ 0 & 0 & 1 \end{bmatrix}$$

$$\begin{array}{c} -0.107 \\ \textcircled{1} \\ -0.107 \end{array} \begin{bmatrix} 4.52 & 0.667 & 0 \\ 0 & 6.28 & 0.667 \\ 0 & 0.667 & 4.52 \end{bmatrix} \begin{bmatrix} 1 & 0 & 0 \\ -0.15 & 1 & 0 \\ 0 & 0 & 1 \end{bmatrix}$$

$$\begin{array}{c} 0.016 \\ -0.15 \\ \textcircled{1} \end{array} \begin{bmatrix} 4.52 & 0 & -0.071 \\ 0 & 6.28 & 0.667 \\ 0 & 0 & 4.45 \end{bmatrix} \begin{bmatrix} 1.016 & -0.107 & 0 \\ -0.15 & 1 & 0 \\ 0.016 & -0.107 & 1 \end{bmatrix}$$

$$\begin{array}{c} 0.225 \\ 0.16 \\ 0.225 \end{array} \begin{bmatrix} 4.52 & 0 & 0 \\ 0 & 6.28 & 0 \\ 0 & 0 & 4.45 \end{bmatrix} \begin{bmatrix} 1.016 & -0.107 & 0.016 \\ -0.150 & 1.016 & -0.150 \\ 0.016 & -0.107 & 1 \end{bmatrix}$$

$$\begin{bmatrix} 1 & 0 & 0 \\ 0 & 1 & 0 \\ 0 & 0 & 1 \end{bmatrix} \begin{bmatrix} 0.225 & -0.024 & 0.035 \\ -0.024 & \cdot 1615 & -0.024 \\ 0.0035 & -0.024 & 0.225 \end{bmatrix}$$

$$\therefore [b_1]\left[[b_1'][f][b_1]\right]^{-1} \times \left[[b_1'][f][b_0]\right] = \frac{1}{2l} \times [b_1] \begin{bmatrix} -0.161 \\ -0.908 \\ +0.135 \end{bmatrix} = \begin{bmatrix} 0 \\ 1.07l \\ 0 \\ -0.322l \\ 0 \\ -1.816l \\ 0 \\ 0.773l \\ 0 \\ 0.27l \\ 0 \\ 0.0805l \\ 0 \\ -0.161l \\ 0 \\ -0.067l \\ 0 \\ 0.135l \\ 0 \\ 0.454l \\ 0 \\ -0.908l \\ 0.454l \\ 0 \\ -0.908l \\ 0 \\ -0.161 \\ -0.135 \\ -0.908 \\ +0.908 \end{bmatrix} \times \frac{1}{2t} \quad \therefore [b] = \begin{bmatrix} 0 \\ 0.93l \\ 0 \\ 0.322l \\ 0 \\ -0.184l \\ 0 \\ -0.773l \\ 0 \\ -0.27l \\ 0 \\ -0.080l \\ 0 \\ 0.161l \\ 0 \\ 0.067l \\ 0 \\ -0.135l \\ -l \\ -0.454l \\ 2l \\ 0.908l \\ -0.454l \\ 0 \\ 0.908l \\ 0 \\ 0.161 \\ -0.135 \\ -1.092 \\ -0.908 \end{bmatrix}$$

$\therefore$ For $R = 10$ tons.

$$[S] = [b][R] = \begin{bmatrix} 0 \\ 4.65 \\ 0 \\ 0.61 \\ 0 \\ -0.92 \\ 0 \\ -3.86 \\ 0 \\ -1.35 \\ 0 \\ -0.40 \\ 0 \\ 0.805 \\ 0 \\ 0.34 \\ 0 \\ -0.67 \\ -5.0 \\ -2.27 \\ 10 \\ 4.54 \\ -2.27 \\ 0 \\ 4.54 \\ 0 \\ 0.805 \\ -0.67 \\ -5.46 \\ -4.54 \end{bmatrix} \begin{array}{ll} a1 & 1 \\ & 2 \\ a2 & 1 \\ & 2 \\ a3 & 1 \\ & 2 \\ a4 & 1 \\ & 2 \\ a5 & 1 \\ & 2 \\ c1 & 1 \\ & 2 \\ c2 & 1 \\ & 2 \\ c3 & 1 \\ & 2 \\ c4 & 1 \\ & 2 \\ d1 & 1 \\ & 2 \\ d2 & 1 \\ & 2 \\ d3 & 1 \\ & 2 \\ d4 & 1 \\ & 2 \\ c1 & \\ c2 & \\ g1 & \\ g2 & \end{array} \quad \text{Members}$$

Fig. 7.37. Matrix manipulation for a Unit Load method analysis

This three redundancy example approaches the maximum degree of complication acceptable for manual computation. Guy Tidbury writing in 'Stress analysis of vehicle structures' (ASAE note one, Cranfield) describes the computer matrix analysis of a slightly less idealized van body structure shown in Fig. 7.38. The effect of a cut out representing a side window is shown, the matrix manipulation being handled by a digital computer.

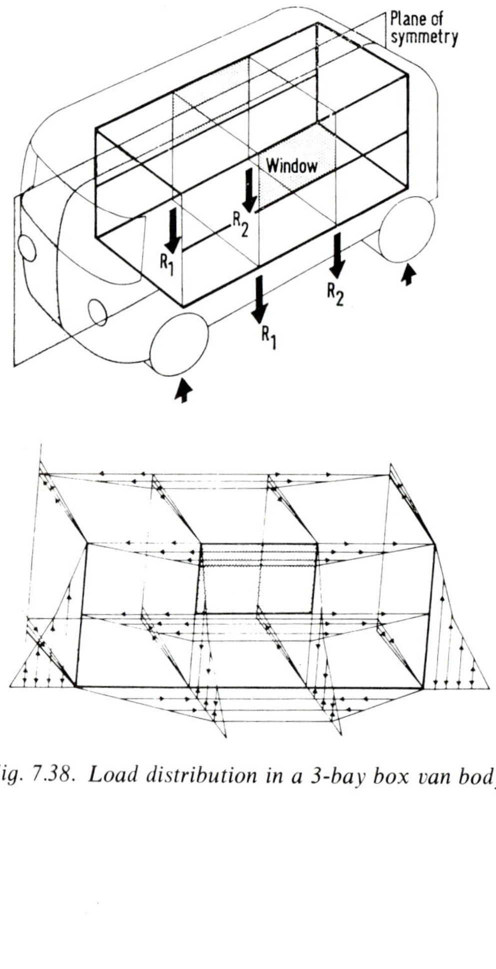

Fig. 7.38. Load distribution in a 3-bay box van body